Our
Blue
Planet

Our
Blue
Planet

THE STORY OF THE EARTH'S EVOLUTION

Heinz Haber

Translated by Ernst Stuhlinger

CHARLES SCRIBNER'S SONS · NEW YORK

Preface

During my long-time activities as a producer of scientific programs for American and German television, members of the audiences often suggested that I publish the text in the form of a book. In television, however, the picture is the principal element; the words alone are inadequate. Then, in the summer of 1964, Radio Munich asked me to present a series of eight lectures on the evolution of the earth. Since a radio talk must convey its full meaning by words alone, that series provided the material for *Our Blue Planet*. The informality of the book's style results from the form of the original presentation.

The book was first published in German in Stuttgart in 1965 and became a best seller in Germany. In the fall of that year, the director of the Second German Television Program, Professor Karl Holzamer, asked me to produce a television series based on its contents. This was broadcast during October–December 1965 in twelve installments, in which the material was covered in greater detail but the main features of the book were retained. The series won the Golden Camera award for German television. Now I am particularly pleased that the book is being published in the United States.

I want to express my appreciation to the Deutsche Verlags-Anstalt for the excellent production of the original edition, and also my special thanks to my wife, Irmgard, who worked faithfully and tirelessly in preparing the manuscript.

HEINZ HABER

Stuttgart

Contents

Introduction

A few years ago, scientific journals published a color photograph that probably left a deep impression on the perceptive minds of many readers. One of a series brought back by Scott Carpenter from his space journey after he had orbited the earth several times, it showed part of the globe as seen from an altitude of about a hundred miles.

Until then, astronomers, meteorologists, and physicists had wondered what the earth would look like, seen from outer space. Even before the Space Age, scientists were certain that the earth would not resemble a geographical globe, despite the popular notion. However, it should be borne in mind that our terrestrial atmosphere is rarely clear and transparent enough so that an observer from space can discern continents, oceans, islands, and lakes as distinctly as shown on a map. More than one half the total surface of the earth is continuously enveloped in dense clouds, and even in areas of clear weather the contours of the continents are not at all as sharply outlined as on a geographical globe. This is why the viewer has difficulty finding his bearings on Scott Carpenter's space pictures, which show land masses as all but obliterated by clouds, with layers that reflect the sunlight with blinding brightness.

The most striking feature in these photographs, however, is the color. Our planet appears enveloped in a brilliant, deep aquamarine blue. Even before the first space flights, geophysicists anticipated that our planet might look blue. Mars appears reddish; Mercury and Saturn yellow; Uranus and Neptune have a greenish tint; and the other planets are white. Our earth is the only blue planet in the solar system.

Another notable quality in the astronaut's pictures is the appearance of the earth's edge, which is curved. It is a wide but obvious segment of a circle, photographic proof of the fact that man is indeed living on a sphere.

Historically, the spherical shape of the earth has played an important role in human thought since the era of exploration began, more than four hundred years ago with the discovery of the American continent by Columbus and the voyage of Magellan around the earth. The development of space travel brought us another big step forward. It made us regard our native planet as one among a large number of celestial bodies, and it helped us to see it in relation and in perspective to its neighbors in space. It is for this reason that now, more than ever before, people think about the nature of our earth as a planet; the notion of the earth as merely a celestial body has vanished. No longer are scientists the only ones who want to know about the earth's origin, how it evolved, how life developed, what forces gave the earth its present shape, and what its future may be.

This book will attempt answers to these questions. Of course, in eight chapters, such a task can hardly be accomplished. Astronomers, physicists, geologists, oceanographers, meteorologists, geographers, glaciologists, chemists, and engineers have accumulated enough knowledge on the subject to fill a library. The broad view to be presented here can only be sketchy. What is offered in this book is based on the present state of scientific knowledge. Not only the earth and life on it, but also the sciences, develop and change. Concepts of the earth's evolution that appeared acceptable and plausible fifty or even twenty years ago are mostly obsolete today; new ideas are bound to replace the older ones. The natural sciences are like a huge crossword puzzle whose rows and columns grow faster than they can be solved. Moreover, for the most part, nature has given us only very scanty hints to the solution of this puzzle. However, this is probably the special challenge of the intellectual adventure called natural science. These thoughts apply particularly to the earth sciences, as to all questions related to the origin and evolution of our planet.

The subject of this book is divided along morphological lines. Each chapter describes one of the phenomena presented by the earth. The first chapter concentrates on the origin of the earth, showing how the earth originated as one among other planets, and that its origin is closely related to that of the sun; also that the formation of planets must be a relatively frequent occurrence in the depths of the universe.

The second chapter deals with the earth's substance, particularly its spherical shape, and the structure of its interior as it has been analyzed by means of seismic waves.

In the third chapter, a completely new idea is discussed, the bold hypothesis of the expansion of the earth. Powerful arguments exist today for the assumption that the gravitational constant has been decreasing very slowly but continuously during the history of the universe, and that it is still decreasing. The gravitational force, however, is that universal force which holds the terrestrial sphere together. If gravity, some billions of years ago, was much greater than it is now, the earth must then have been much smaller. With the slow decrease of the gravitational force during the geological evolution, the earth must have expanded slowly until it reached its present size. Obviously this hypothesis requires that present-day views of the evolution of the earth must be revised in many respects. This revision has already resulted in a number of very interesting ideas about the origin of continents and the development of climate.

The subject of the fourth chapter is the age of the earth. It is impossible to develop ideas about the evolution of our planet without considering those inconceivable eons during which this evolution occurred. It has been common knowledge for some time that the earth must be much older than previously assumed—older, for example, than the seventeenth-century Irish archbishop James Ussher's calculation of six thousand years, plus or minus one day. Modern scientific methods are described which permit a highly accurate determination of the earth's age.

Chapter 5 deals with the origin of the atmosphere and the sea. The earth is the only planet in our solar system with a worldwide expanse of ocean, and with an atmosphere which contains as much as 20 percent free oxygen. How does it happen that ours is the only planet having the two attributes that are so necessary for life?

Life and its origin is the subject of the sixth chapter. In this regard, too, the earth seems to hold a privileged position. It is likely that only Mars may bear some primitive forms of plant life, whereas life on earth has developed a proliferation of forms. The reactive effect of life's existence, which exerted so great an influence on the development of the earth, is also described. It should be noted, in contrast to earlier theories, that the phenomenon of life does not wait patiently until nature's inanimate processes have created conditions whereby life can exist in all its various forms. Rather, the conclusion has been reached that life with its countless chemical processes acts in a decisive way to generate its own environmental conditions, transforming a planet in such a fashion that it achieves the capacity to support life in ever greater variety.

Chapter 7 deals with climate, which is defined as the sum of all environmental conditions under which life, particularly a species such as man, is extant. Life must continuously cope with its climatological environment. The climate on earth has by no means always been the same as now. Remember, for example, that there have been several Ice Ages.

The last chapter discusses the future of the earth—how the earth tends to develop as an abode of man, and whether or not the existence of mankind itself may be threatened by natural catastrophes.

1 ❋ Origin of the Earth

An inquiry into the origin of our planet may best be started by taking a closer look at the scene of action where this gigantic development occurred: the universe. The universe is an enormous, three-dimensional space whose extent exceeds our ability to comprehend it. Any attempt to compare its magnitude with anything known is vain. The space throughout the universe is inhabited by stars which, in size and luminosity, are similar to our sun. A sense of the vastness of the universe may be obtained by considering the sun. Seen from the earth, the sun is an object of overwhelming, blinding brightness. Remember that all the minute specks of light in the starry sky are suns like our own, and imagine how far-reaching the universe must be! If we look at the sky on a clear night, we can see about three thousand stars in our hemisphere; most of these exceed our sun in absolute luminosity. All the individual stars we can discern from the earth belong to our galaxy, the Milky Way.

Compared to solar diameters, the mutual distances between the suns of the Milky Way are far, far greater. A scale model of part of the universe in the vicinity of the sun, reduced to the size of the United States, would show spheres the size of a walnut at all the locations of capital cities throughout the states. The space between the walnuts would be virtually empty.

Planets are celestial bodies of the second order, orbiting their suns; their diameters are twenty to a hundred times smaller than solar diameters. Referring again to the scale model, the planets' orbits cover an area of only a block or two around the walnut-sized suns. In fact, the planets in the scale are only dust particles—pinheads at best—in close proximity to their central bodies. The countless suns are not distributed at all uniformly over the total volume of the universe. They are concentrated in huge star systems, of which our Milky Way is an example. The Milky Way contains about two hundred billion suns, all similar to our sun in magnitude and luminosity; they have all accumulated at one specific place in the universe. In visualizing the structure of the Milky Way, bear in mind that all these individual suns are separated by those enormous distances which the scale model simulated by the distances between walnuts in each of the American capitals. Viewed from where we stand, however, the two hundred billion suns of our Milky Way appear as a uniform, relatively compact system of stars, or rather a system of star clouds arranged in a flat spiral.

A star system like our Milky Way is not a rare occurrence in the universe. There are hundreds of billions of other Milky Way systems, each about as large as our own, each consisting of hundreds of billions of stars. These other Milky Ways or galaxies are at various enormous distances from us. Yet many are so near, comparatively, that they can be photographed with large telescopes; they look like faintly luminous spirals. The spiral nebula in Andromeda, one of the galaxies nearest to the Milky Way, is two million light-years distant. The farthest galaxy photographed thus far is several billion light-years away.

Speaking of the earth, which is our main subject, it is well to remember that it is a mere speck in the universe, belonging to one sun in a galaxy of hundreds of billions of suns, and to a single galaxy among hundreds of billions of others.

According to recent concepts of astronomy, the origin of the earth cannot be separated from the origin of the sun. Science has established with a high degree of probability the fact that the earth and the other planets are about the same age as the sun. Therefore, the origin of the earth cannot be discussed without considering also those forces and processes which have led, and are still leading, to the formation of stars.

Students of morphology are surprised and impressed by the fact that the forces of nature have created a large variety of individual entities, each of them with a great number of representatives. To this end, nature had at its disposal the matter which composes the universe, spread out over space and time. Nature must have proceeded according to a specific plan in using this matter for individual creations: galaxies, stars, planets and meteoroids, giant sequoia trees, algae and spores, elephants, mice, flies, and human beings. However large the number of individual entities may be, the enormous quantity of atoms within each particular entity is equally remarkable.

Modern scientists have serious reservations about discussing the fundamental act of creation; they do not feel competent to do so. For this very reason, this book begins with the second chapter of creation, assuming that developments have already proceeded to a point at which space, time, and matter already exist. The concept of matter will be described in more detail. Today's physicists find it quite tempting to assume that in its original state matter consisted only of one kind of basic building blocks: hydrogen atoms. Imagine that the volume of the universe was filled with an exceedingly rarefied gas consisting only of hydrogen atoms. A fundamental force was exerted between these particles—the force of gravity. In time, this force led to concentrations of hydrogen gas within the huge volume; centers of condensation formed wherein many hundred thousand billion tons of hydrogen were concentrated. The closer the hydrogen atoms moved together, the greater became the force of their mutual attraction, resulting finally in huge spheres of dense gas. This act of creation has occurred an uncounted number of times in the history of the universe, and it will go on doing so in time beyond measure.

The physics of such a sphere of gas cannot be discussed in detail. Suffice it to say that in the center of the sphere extreme pressures and extreme temperatures develop, which lead to the gradual transformation of the hydrogen into helium and, during later phases of stellar evolution, into heavier elements such as oxygen, iron, gold, and uranium. Most of these transformations result in the release of energy that is emitted into space in the form of electromagnetic radiation. It has become a common belief among scientists that the interior of stars during stellar evolution has been, and still is, the birthplace of the whole range of chemical elements.

For some decades astronomers by means of powerful telescopes have been able to observe from time to time a gigantic star explosion in one of the nearer galaxies. It is believed that these explosions are spreading the chemical elements generated within the stars over vast reaches of the universe, and that these elements, again in the form of gas and dust clouds, are providing the material for the formation of stars of a second, third, and fourth generation. This process can be likened to that by which the seed pods of a plant explode and scatter their contents all around, thus initiating the growth of the next generation.

The matter which composes the sun and the planets today contains a small but significant portion of heavy chemical elements. This fact implies that the sun and all its planets, including the earth, must be members of the second, of the third, or perhaps even of a later generation in the creation of the universe.

In earlier times, it was held that the planetary system owed its existence to a kind of birth from the body of the sun. This rather primitive concept had to be abandoned when it was found that the sun, on the one hand, and planets Mercury, Venus, earth, and Mars on the other, have altogether different chemical compositions. The sun consists largely of hydrogen and helium, the two simplest, lightest, and most basic chemical elements. The earthlike planets, however, consist mostly of the heavier elements, oxygen, silicon, iron, nickel, and many others. The main ingredients of the sun, hydrogen and helium, are quite rare on the terrestrial planets. This is why those early, naïve concepts about the origin of the planets had to be discarded. The formation of the planets cannot be explained by simply assuming that part of the solar matter, when scooped out of the sun and thrown into space, would cool down and form a planet. This would be as unlikely as the formation of a lump of steel out of a vat of dough. When the sun and the planets took form, a number of specific processes must have occurred which concentrated the heavier elements that make up only about one percent of the solar matter, and which finally made these heavier elements condense into planets.

These ideas are the basis of a theory of the origin of the planetary system which the German astrophysicist Carl-Friedrich von Weizsäcker developed in 1938. His thoughts have found widest acceptance; they were further refined and worked out in great detail by other physicists and astronomers during the past twenty-five years. Weizsäcker's theory assumes that the sun and the planets originated from a huge cloud of gas, of which 99 percent consisted of hydrogen and helium, and only 1 percent of the heavier elements, oxygen, silicon, iron, nickel, and others. This gaseous cloud condensed under the influence of gravity in the course of millions of years. About 90 percent of all the material in the cloud concentrated in a central body, the proto-sun, while 10 percent was left over as the far-reaching atmosphere of the proto-sun. One hundredth of this 10 percent consisted of heavy elements.

In the course of time, a system of regularly arranged vortices formed in this primordial solar atmosphere; the gaseous matter condensed at those places where we find the planetary orbits today. It is the decisive idea of this theory that one hundredth of the 10 percent of the total mass of this system, or one thousandth of the total mass of the sun and the planets, was the heavy elements.

As contrasted to hydrogen and helium, the heavy elements are capable of forming particles and grains on the basis of chemical forces. They were brought together by gravitational forces, coagulated, and finally formed the planets. These processes can still be recognized in the abundance of chemical elements in our solar system. The primordial, far-reaching solar atmosphere from which the planets originated had the same composition; this atmosphere represented about 10 percent of the solar mass. The total mass of the planets amounts indeed to one thousandth of the solar mass, so the account checks.

Our earth is the product of a very intriguing process of selection which nature applied in the creation of the planets. The earth is indeed a most precious body! It consists of a concentration of extremely rare and valuable heavy elements, without which the great variety of phenomena on the earth would be unthinkable. Using hydrogen and helium, even nature can create nothing but a sun in the shape of a gigantic glowing ball. When the ball is big enough, temperature and pressure in its interior rise to such a tremendous degree that the sphere begins to radiate fiercely. This, however, is all that can happen in this early phase.

In the planets, on the other hand, nature has accumulated the extremely rare and, as seen in a cosmic view, extremely precious heavy elements. In doing so, it prepared the chemical and physical stage on which the multicolored drama of planetary evolution can take place. The other eight planets, of course, came into existence simultaneously with the earth. They, too, consist mostly of those rare heavy elements.

This current theory about the origin of the planets was able to explain also another very strange phenomenon in the arrangement of the planetary orbits, which had been a puzzle for a long time. It is well known that the diameters of the planetary orbits increase from planet to planet with remarkable regularity, a feature that has even been represented by a simple numerical series. It is a necessary consequence of the processes already described that the vortices in the primordial solar atmosphere had to assume larger and larger dimensions, the farther out they were. The laws of motion of these vortices provided almost automatically the distances at which planets could originate. Indeed, the nine planets of the solar system orbit the sun at these distances.

These facts permit a further conclusion which is, from a philosophical viewpoint, extremely interesting and consequential. What applies to the sun must also apply to a number of other stars that are similar to our sun in size and structure. In former times, it was assumed that a phenomenon of the complexity and precision of our planetary system must be exceedingly rare; however, more recently the conclusion was reached that numerous systems of this kind must be scattered throughout the universe. During the detailed discussion of the evolution of our planet in the following chapters, the reader should keep in mind that the same evolution has occurred on countless other earthlike planets at innumerable other places in the universe.

2 ⁕ Internal Structure

The actual formation of the earth took a relatively short period of time, something like a hundred million years. Periods of millions of years are really brief time spans, considered on a cosmic scale. In all likelihood, the earth is several billion years old; its formation, therefore, covers just a small fraction of its total age. On an abbreviated scale, in which the earth's age is shortened to one year, the formation of the earth would have required about a week, and this is short indeed!

In principle, ingredients of a planet such as our earth are simple: take about two and a half thousand billion billion tons of iron and nickel, an equal quantity of oxygen and silicon, a little less aluminum and magnesium, add a trace each of the remaining chemical elements, and mix. This mass of material develops an enormous gravitational force which tries to pull all the parts as close together as possible. The entire mass, therefore, assumes the shape of a sphere, with a diameter of about 8000 miles. In case this big ball happens to be in the vicinity of a G-type sun, and if it happens to orbit this sun at a distance of a hundred million miles, it will eventually become an earth. Nature used this recipe when it created the earth. The necessary ingredients existed in the form of gas, dust particles, and larger chunks of solid matter in the primordial solar atmosphere. In the course of time, this material was drawn together by gravitational forces, whereby it assumed the shape of a sphere; these forces, exerting their influence equally on all the parts of the system, result in an even distribution of matter around the center. Thus a body the size of the earth can exist only as a sphere. As parts of the earth first moved together by the action of gravity, most of their kinetic energies were transformed into heat. The ball of matter therefore became hotter and hotter. Extremely high temperatures developed in the interior of the sphere, while the surface cooled off through radiation.

The deeper the penetration toward the earth's center, the higher the temperature. This is proven not only by hot springs and volcanoes, which produce boiling water and white-hot, molten lava; it is evident also in mine shafts and drill holes. Precise measurements have shown that with each mile a probe penetrates into the earth, the temperature rises by about 85° F. Two miles below the surface, the temperature of boiling water is reached, and at twenty-five to thirty miles, the heat rises to 2000° to 2500° F. At this temperature, the rocks become molten. The temperature at the center of the earth is an estimated 6500° F. Internally, the earth glows white hot; fortunately, its thin shell is cool enough so that we do not burn our feet. Because of this high temperature at the center of the earth, our planet contains a tremendous amount of heat energy, and the question arises where this energy came from. The bits and pieces that formed the earth a long time ago were relatively cool. In fact, their temperatures were not much higher than that of the earth's present surface, since they orbited the sun just as the earth does. When, however, those meteoritic particles fell onto the growing planet during the formation of the earth, their kinetic energies generated a temperature of several thousand degrees Fahrenheit under the conditions then existing. The processes that led to the formation of the earth eons ago have not ceased even now. Daily, the mass of the earth increases by about a thousand tons, because of the meteoroids and meteoritic dust particles that strike the earth every twenty-four hours. On a clear night, the luminous paths of these meteors can be readily observed, conveying some idea of the tremendous energy and resultant heat given off by each of these "shooting stars." So great is their velocity as they enter the earth's atmosphere that the friction ignites the particles to an enormous intensity of heat, which is the reason for their blazing brightness.

Ever since the earliest beginnings of the earth, the major part of its mass has been extremely hot. The heat is retained by the outer crust which protects the body of the earth against heat loss, like the wall of a thermos bottle. There are even reasons to assume that the temperature of the body of the earth increased, rather than decreased, during the several billion years of its existence. The earth contains radioactive elements, particularly in its outermost layers, which slowly but continuously emit radioactive energy. This energy, transformed into heat, is retained over long periods of time. Sometimes a channel opens from a hot region to the surface; a volcano erupts, and a broad stream of red-hot lava flows down the slope of the mountain, often engulfing fields, houses, and entire villages. The temperature of this molten lava is higher than 2000° F., which is the temperature within the earth at a depth of about twenty to thirty miles. One might believe, therefore, that the entire core of the earth, except for a thin crust, is in a liquid state. This, however, is not so. When considering the state of the material in the interior of the earth, it must not be forgotten that this substance is exposed to an extremely high pressure. At 2000° F., lava flows like syrup; it can be rightly called a red-hot liquid. But this is true only under the conditions prevailing at the earth's surface where the lava is not exposed to pressure. At a depth of thirty miles, where the temperature of the material is more than 2000° F., the pressure exceeds 20,000 atmospheres. Under such conditions, the lava behaves entirely differently, assuming a state which the physicist calls "plastic."

A good example of plastic material is sealing wax. A stick of sealing wax is solid; at room temperature it is relatively hard and brittle. If a piece were put on an anvil and hit with a sledge hammer, it would splinter like glass. However, placed in a cup and left alone for several years, it would gradually lose its original shape, eventually filling the bottom of the cup as cleanly and smoothly as a residue of coffee. Such is the property of plastic substances. Evidently they behave like solids, yet over long periods of time they seem to act like liquids. This is the way in which the structure of matter in the interior of the earth must be visualized.

This little detour with the sealing wax was necessary to make the development and the structure of the core of the earth understandable. As the material that accumulated during the formation of the earth reached higher and higher temperatures, it became all the more compressed by the mounting pressure of the ever-growing mass. Seen at a glance, the body of the earth is entirely solid; observed over long periods of time, it is liquid. Earlier in the evolution of our planet, the various materials which compose the earth underwent a stratification. An extremely slow movement of the material in the interior of the earth occurred, far slower than the movement of the sealing wax in the coffee cup. However, nature takes its time—and time it has. Those processes which did not occur during the first hundred million years after the birth of the earth simply occurred during the second, third, fourth, or fifth hundred million years. In essence, the event that took place was similar to what can be observed in an ore smelter: the heavy metals sink to the bottom, and the lighter slag floats on top. The bottom in the spherical earth, of course, is the center. In this way the various materials in the earth have separated in the course of these inconceivably long periods of time. As a result of this process, our planet is structured like an onion, with concentric shells.

Beginning at the top, the first layer consists of the lightest material—the earth's crust, only twenty to thirty miles deep, in which granite is the dominating substance. It has cooled and solidified. The next layer is a thin stratum of basalt, heavier than granite; the basalt layer also is part of the crust. The next layer is the mantle of the earth, a thick shell consisting mostly of the mineral olivine. Olivine is a heavy iron-magnesium silicate. The crust represents merely a thin skin, comparatively like the shell of an egg; the mantle, however, extends almost 2000 miles in depth —a thick shell containing more than two thirds the mass of the earth. The central part, finally, is the core with a diameter of about 4000 miles. Presumably it is an extremely dense, hot mass of iron and nickel. Admittedly, little is known about the core of the earth. There are scientists who hold that a large portion of the earth's core consists of hydrogen in a special physical state.

How can all these details be discovered? After all, the deepest drill holes have not reached farther than about four miles. No more than a fifth of the earth's outer crust has been penetrated. The mantle, to say nothing of the core, will be more difficult to reach than the most distant planet in the solar system.

Whatever is now known about the structure of the earth's interior has been deduced from the analysis of the propagation of seismic waves occurring as a result of earthquakes.

An earthquake is a more or less violent concussion originating in a small and well-defined center. Earthquakes originate usually at a depth of six to sixty miles. They are caused by stresses in the top layers of the earth which build up over periods of years, decades, or centuries, and which are relieved in violent spasms. The material of the earth's crust behaves somewhat like the wood in wall paneling. In a wooden wall, surface stresses sometimes occur which are then relieved with audible cracks. An earthquake is a similar event, transposed to the gigantic dimensions of the earth. From the local center of concussion, seismic waves propagate in all directions. These waves move rapidly, traveling through the earth almost as if through a sphere of solid steel. When a tremor occurs, the earth reacts like a bell hit by a clapper. If one touches a ringing bell with the fingertips, the vibrations are felt distinctly. The same effect is seen on the earth. Seismic waves originate in the center of the earthquake and move out radially, some traveling along the surface of the planet, others spreading downward and penetrating the entire body of the earth.

Sensitive instruments, called seismographs, have been developed to record these concussions even at distances of thousands of miles from the center of origin. Earthquakes by the thousand occur annually. Geophysicists have recorded so many shocks since the invention of the seismograph that a fairly accurate knowledge exists today of the various modes of propagation of earthquake waves. The seismic waves that start at the center and penetrate the depth of the earth do not follow straight lines at all. The deeper they penetrate into the earth, the denser the material through which they must travel, and in doing so they are refracted and reflected in accordance with physical laws. The curved and broken paths which the waves follow permit the physicist to draw conclusions about the density of the material through which the waves pass. One of the most important objectives in these observations is the behavior of the waves when they encounter an interface in the earth's interior. When the seismic waves arrive at the interfaces between the earth layers, they are partially reflected, like light rays in a mirror. Careful analysis of these reflections led to the discovery of the shell-like structure of the earth and even made it possible to determine the thickness of each layer.

Comparing the earth with an onion is not quite correct, because an onion consists of a large number of similar layers. A better comparison would be a hen's egg, more or less spherical. With this as a model, the core of the earth corresponds to the yolk, the mantle to the egg white, and the crust to the shell. The model offers the added advantage that the relative dimensions of its spherical components coincide with those of the earth. Comparatively speaking, the core of the earth is about the size of the yolk, the mantle is about as thick as the egg white, and the crust is much like the shell.

Man lives on the earth's surface; he therefore knows best the outermost layer or crust or, referring to the model, the shell. The crust of the earth is immediately accessible to the probing and exploring mind of man. We can dig into it, and it can be explored with the tools of the geological sciences. Thus considerable knowledge about the structure and composition of the crust has been acquired. As mentioned earlier, the crust of the earth consists primarily of two different materials, one lighter than the other. The upper layer is granite, the lower basalt. In a chemical sense, the upper layer is composed mainly of silicon and aluminum, whereas in the lower and heavier layer silicon and magnesium are predominant. For this reason, geologists and geochemists have coined special names for these layers. The upper, lighter layer is "sial," which is derived from silicon and aluminum. The lower is called "sima," from the first syllables of silicon and magnesium.

The continents, or the dry land, consist mainly of sial, while the bottom of the ocean is made up largely of sima layers. It is a strange fact that, at the places of the continents, the crust is almost twice as thick as in the regions covered by oceans. The continental crustal blocks—and that is why they were given this appropriate designation—are gigantic slabs of light material that float on the heavy, plastic material of the mantle. A continent may be compared with a thin cracker floating on thick pea soup. In this model, the heavier sima layer must be compared to the surface of the pea soup which has begun to dry up and to form a kind of skin. This dried-up skin represents the wide areas of the ocean bottom as well as the bases of the continental crustal blocks; in our model, it is the base for the cracker floating on the soup. It has long been known that the theory illustrated with pea soup and cracker is borne out by numerous observations. For example, it has been observed that the northern parts of the continents, such as Scandinavia in Europe, and Canada on the American continent, rise gradually. Along the coast of Sweden and Norway, rocks have been found that today protrude by thirty, forty, or more feet above the sea, and which still show distinct marks of an ancient coast line.

This phenomenon is easily explained. About a hundred thousand years ago, during the later phases of the Ice Age, the northern regions of the continents were buried under a layer of ice about half a mile thick, whose enormous weight, borne for several hundred thousand years, pushed them into the soft material of the mantle. When, about twenty-five thousand years ago, the last glaciation came to a close and the gigantic glaciers melted away, the continents were relieved of their icy burden. The situation is like that of a heavily laden ship afloat in a sea of tar and suddenly unloaded. It should be remembered that the material of the mantle, on which the continents float like boats, is plastic—not liquid like water, but a viscous substance. When the continents lost their burden of ice at the end of the last Ice Age, it took thousands of years before they rose by an appreciable amount. This process is still going on today.

Our knowledge about the interior structure of the earth comes mainly, or almost completely, from studies of the propagation of seismic waves. It is understandable, therefore, that geophysicists have eagerly sought other means and methods to confirm, to expand, or even to revise these concepts.

The fact that the exploration of the earth's interior can be supported and greatly advanced by the modern technology of space flight should serve as an impressive example of the close correlation between different areas of scientific endeavor. Such correlation between earth science and space exploration seems at first glance far removed, because space flight tends to focus our attention away from the earth, in an opposite direction from the subject under discussion. Yet it is a fact that an artificial satellite, orbiting the earth at a distance of several hundred miles, is able to provide detailed information about the internal structure of the earth.

In order to understand this, it should be remembered that an artificial earth satellite is kept in its orbit by the gravitational force of our planet. Generally the orbit of a small celestial body around a large central body, which can be calculated with the laws of celestial mechanics, is determined primarily by the mass of the central body. In principle, the laws of physics needed for the calculation of the orbit of a satellite are simple, and the mathematical equations which result can be understood by any intelligent high-school student. It must be admitted, though, that these equations are based on a fundamental simplification, namely, the assumption that the entire mass of the central body is concentrated in one point at its center. The simple equations representing the orbit of a satellite actually contain this simplification.

The discoverer of the law of gravitation, the English scientist

Isaac Newton (1642–1727), who was the first to establish the equations for the calculation of the orbits of the moon and the planets, introduced this simplification. Newton was a young man when he discovered the law of gravitation, but he waited several decades before he published this fundamental discovery. The reason for his long hesitation was his desire first to prove the justification for this simplifying assumption. He finally succeeded in showing that the gravitational field of a sphere of completely uniform density and hence completely homogeneous structure acts exactly as if the total mass of the sphere were concentrated, in a mathematical sense, in its center. Having provided this proof, Newton could then successfully apply his famous law in this simplified form.

The body of the earth does not have a completely uniform, homogeneous structure throughout its interior. In the center there is the heavy core; then follows the lighter mantle; and finally, the still lighter crust forms the outermost layer. Besides, the crust, because of the distribution of the continents and oceans, has a highly variable thickness. Furthermore, the earth is by no means a true sphere. Its surface is uneven, and its spherical shape is flattened at the poles as a consequence of its rotation around its axis. The result is a gravitational field which is not at all as uniform and symmetrical as the field of a mathematically perfect sphere. A little earth satellite, which is controlled in its orbit around the planet in a very sensitive manner by the gravitational field of the earth, will indicate by its own movements any irregularities of the gravitational field. This has actually been observed by precise analyses of the minute variations in the orbit of an artificial satellite. The origin of these irregularities must be sought in the irregular structure of the earth's interior. Although these studies are not yet complete, it is already obvious that the modern technologies of space flight can help substantially in the study of the inner structure of the earth. It should be noted that the first results of this most interesting method of research are in full agreement with earlier views about the essential features of the earth's internal structure.

As residents of the surface of the earth, scientists have a considerable amount of information about the structure of the earth's crust. There is no doubt that the division into water and land, or rather into continental crustal blocks and ocean bottom, is a surprising circumstance whose implications have long been overlooked. Only recently, a fascinating hypothesis provided a clue to the interpretation of this strange phenomenon—the hypothesis of the expansion of the earth.

3 ✳ Expansion of the Earth

The most striking structural feature of the surface of our planet is the separation of water and land. This situation brings up the question of how the continental crustal blocks came into existence. The modern hypothesis of the expansion of the earth offers an elegant explanation of this process.

In comparing the earth with its neighbors within the planetary system, it is safe to say, without running the risk of seeming parochial, that our planet is unique. In fact, it is the only one with a world-wide ocean. All the others, along with their moons, including our own satellite, are covered either by absolutely dry desert or by vast layers of ice and frozen methane and ammonia. Were it not for a peculiarity in the earth's evolution, its entire surface would consist of water extending from pole to pole, without a speck of land. If the earth were a smooth sphere, the quantity of water in the oceans would be sufficient to envelop the entire globe with a depth of 7500 feet, in which case this planet would have a truly liquid surface.

Fortunately for humankind and for all land-living animals, the oceans are limited to 71 percent of the earth's area—still a considerable portion. It should be noted that the continents, islands, and even the ice-inundated polar caps occupy less than a third of the surface.

It was pointed out earlier that the earth is structured like an egg, although it is not that shape, actually, but spherical. In the center, there is the core, corresponding to the yolk, representing the densest part. Then follows the mantle, corresponding to the egg white. Presumably the mantle consists mainly of the heavy mineral olivine. Finally, there is the shell, corresponding to the crust. The crust of the earth floats on the plastic material of the mantle like a kind of clinker. At this point, however, the similarity between the earth's crust and an egg shell comes to an end; in contrast to an egg shell, the crust consists of two superimposed layers of different densities, and besides, unlike an egg shell, the crust of the earth does not have the same thickness throughout.

Geologists have come to the conclusion that the continental crustal blocks are huge slabs of granite, ten to twenty miles thick, which rest on a thin layer of basalt three to six miles deep. The floor of the oceans, however, consists merely of a thin layer of sediments a few miles thick. These sediments were deposited directly on the layer of heavy basalt. It is known that the earth's crust is thinnest at the bottom of the sea, and that the continental crustal blocks are thicker portions of the crust that float in the plastic material of the mantle like icebergs; they protrude enough to reach levels high above the sea. The continental crustal blocks reach deepest into the mantle beneath the high mountain ranges. Like a heavily laden ship, the crustal blocks have a considerable draft at these places.

The definite division into water and land, mentioned previously, is a phenomenon by no means restricted to the uppermost surface of the earth. In a morphological sense, it reaches deep into the body of the earth, and it is also reflected by noticeable differences in the chemical structure of the continents as compared to that of the ocean floor. In fact, this division characterizes in a typical manner the physics, the chemistry, the dynamics, and very particularly also the morphology of the earth's surface.

Many years ago, the German geophysicist Alfred Wegener (1880–1931) observed that a strange situation exists in the distribution of altitudes on the surface of the solid crust of the earth. If the distribution of altitudes as a function of the frequency of their occurrences is plotted graphically, it is obvious that two altitude levels occur far more frequently than others. Admittedly, there is an altitude difference of about twelve miles between the highest peaks of the mountain ranges and the deepest trenches in the bottom of the sea. However, great altitudes and extreme depths are rare, and they occur only on a very small part of the surface of the earth. By far the greatest part of the earth's surface is either at a depth of about three miles below sea level, or at an altitude of about a thousand feet above sea level. The underwater level is represented by the ocean floor which extends over millions of square miles at an average depth of about three miles. It is mostly flat, but it is interspersed at many places by submarine mountain ranges and deep sea trenches. The altitude level of about a thousand feet above the sea is found in the wide plains of all continents; it is interrupted only to a small degree by hills, and even less frequently by high mountains. This surprising partition into two altitude levels has been known for a number of decades, but it was always regarded as incidental. However, justice would not be done to the subject of this book if the circumstance were left without an explanation.

Without this partition, there would be no dry land, because the earth would be covered from pole to pole by one huge, worldwide ocean.

Alfred Wegener made another interesting observation in reference to the striking similarity in the contours of the west coast of Africa and the east coast of South America. When these continents are drawn on paper and cut out, they can be fitted together almost without leaving a gap; the part of South America that extends far to the east, from the mouths of the Amazon River to the city of Recife, fits into the Gulf of Guinea at the west coast of Africa. It has been pointed out that this coincidence is not really exact, in spite of a relatively fair agreement in the main features. However, critics may have made a mistake by referring to the present geographic coast lines of the two continents. This would not be correct from the morphological, the physical, or the geological standpoint. The coast lines of the continents, as we know them today, are largely the result of chance, depending upon the level of the sea. If the sea level rose or dropped by only a few hundred feet, the geographic contours of the continents would change altogether, even to the extent that some continents would no longer be recognizable.

If one wishes to determine the true boundaries of the continental crustal blocks, which are independent of the varying water level of the sea, one must reckon with the contours of the so-called continental shelf. It is known that each continent is surrounded by a kind of border below the waterline. Along this shelf, the oceans are only a few hundred feet deep; farther away from the coast, the continental blocks drop steeply toward the floor of the deep sea. This is the place where, in a geometrical sense, the transition occurs from the altitude level of the continents to the altitude level of the ocean bottom. The contours of the continental shelves are the true boundaries of the continental crustal blocks. The Australian geophysicist S. Warren Carey has shown that the contours of the continental shelves of Africa and South America indeed match each other so closely that a fortuitous fit must be ruled out.

A further argument is the fact that the geological formations at the African west coast and the South American east coast are the same at corresponding places. Actually, there is not much room left for doubt that these two continents, which are widely separated today by the Atlantic Ocean, were one huge, single structure eons ago.

This is the hypothesis which Wegener established several decades ago as part of his famous theory of the continental drift.

In Wegener's concept, all the continents of the earth at the beginning formed one continuous proto-continent that covered as a single huge island about a third of earth's surface; the rest was ocean. Geological forces, so he assumed, broke up this proto-continent into different pieces in the course of time. The pieces were driven apart by centrifugal forces generated by the rotation of the earth, until finally they settled in the relatively uniform distribution as seen today.

The idea was ingenious; it shook the scientific world and has provided food for thought among geophysicists and geologists ever since. However, it could be shown that the centrifugal forces generated by the rotation of the earth are too minute to tear up a proto-continent and drive the fragments apart. This material is so tough that far greater forces are necessary to move a continental block. This was such an effective argument that Wegener's attractive theory had to be abandoned.

In spite of the rejection of Wegener's theory, the fact remains that the contours of the continents do fit remarkably well, as in the case of Africa and South America. Also, there is this equally surprising partition in the frequency of altitude levels on the surface of the earth, for which a good explanation has not been given as yet. The German physicist Pascual Jordan has presented a bold hypothesis, called "The Expansion of the Earth," which not only offers a solution to these two unresolved questions about the evolution of our planet, but also explains several other strange phenomena regarding the earth's structure. Professor Jordan points out that in all likelihood the earth is in a process of continuous, slow expansion. This novel concept, in sharp contrast to the classical picture of the shrinking apple, has captured the interest of a large number of leading scientists who have recognized that many features of the earth's morphology can be better explained if one does not assume a continuous, slow contraction of the earth, but rather the contrary. However, some scientists place less emphasis on this theory than I and others do.

The hypothesis is so revolutionary, so opposed to all previous concepts, that it is first necessary to explain how such an expansion could arise, and where the physical forces for this gigantic process could originate.

The earth was formed by the accretion of dust particles and gas, which came together because of gravity. As a result, the total material that now constitutes the earth contracted into a sphere. So great is the gravitational force that at the bottom of the thin crust of the earth the pressure amounts to 20,000 atmospheres. Toward the center of the earth, the pressure rises to a value of about 3 million atmospheres, a force equivalent to 20,000 tons per square inch at the earth's center. At first it may seem inconceivable that there could be a physical force strong enough to overcome this enormous pressure throughout the terrestrial sphere and to cause the earth to expand like a rubber balloon being inflated. This difficulty, of course, was obvious to physicists and geologists when they first considered the expansion hypothesis. For that reason, they chose a decisive factor in their argument: the basic nature of gravity.

Gravity is responsible for the contraction of the material which forms the earth, and it is therefore also responsible for its size. The greater the force of gravity, the more the material of the earth will be compressed and the smaller will be the sphere. If for some reason the gravitational constant suddenly doubled, the weight of the layers upon the core of the earth would also double, and the earth would shrink with a catastrophic shock.

The opposite is also true; if the gravitational constant diminished, the material within the earth would be partly relieved of the burden of the overlying masses, and the earth as a whole would expand beyond its present size. Ever since the action of gravity was revealed by Newton some three hundred years ago, it had been accepted as an axiomatic fact that the magnitude of this force is a fundamental constant of the universe, and that this constant never changes, just as the velocity of light and the mass of the electron are assumed to be absolute constants. Actually, this is why such fundamental physical quantities are called "constants of nature."

In the early 1930's, however, the English physicist and Nobel Prize laureate Paul A. M. Dirac presented arguments for the possibility that the gravitational constant has been decreasing continuously, although extremely slowly, in the course of the history of the universe.

What does Dirac's hypothesis tell us? Imagine two steel plates, each of them an inch thick and weighing 400 tons, which float parallel to each other in the universe at a distance of one inch. Under these conditions the steel plates attract each other with a force of one ounce. This force is the result of the general gravitational attraction of masses found everywhere in the universe. The example offers a measure of the magnitude of this force. An understanding of this force can be obtained with the help of a mental experiment. Suppose a tiny helical spring were put between the steel plates; the spring would be compressed somewhat as a consequence of the attractive force between the two plates. The degree of compression serves as a measure of the force; it would be the same if the spring were put vertically on a table at the surface of the earth and loaded with a one-ounce weight.

According to Dirac's hypothesis, this mental experiment would have led to a different result if it had been made with the identical spring but at different periods during the history of the universe. Dirac assumes that four or five billion years ago the attractive force between the two steel plates was greater than at present, and that the spring between the two plates, as well as the spring on the table under the one-ounce weight, would have been compressed even further. In the course of four or five billion years, the attractive force decreased continuously, but extremely slowly, down to the value observed today. In the future, also, the gradual weakening of gravitational force will continue.

Today, the attractive force between the two steel plates in the universe, and also the force exerted by the weight of one ounce of mass, lead to the compression of the spring by a well-defined quantity. However, several billion years ago, the force of attraction was greater, and the compression of the spring would have been about twice as much as today. This example shows that the concept of the British physicist has a fundamental character, and that it penetrates deep into the intrinsic essence of nature. Perhaps one of the most interesting aspects of modern science is that it forces us to search for such fundamentally new concepts, and to visualize variations that are so slow as to become noticeable only in the course of millions of years. Actually, it is necessary to consider such possibilities if the development of our planet is to be understood.

Obviously a slow decrease of gravitational force would have influenced the earth in a decisive way. Assuming this concept to be correct, the earth was much smaller at the time of its birth than it is today. Ever since the earth came into being, the gravitational force has decreased slowly, and the earth expanded at the same rate. The earth keeps growing continuously, so slowly, however, that the increase of its diameter during the entire history of man is no more than a few feet. It should be remembered, though, that the history of *Homo sapiens* covers only a million years, or thereabouts, whereas the earth is several billion years old.

Assume that the earth's diameter was only 4000 miles several billion years ago; the magnitude of the gravitational force was probably large enough at that time to compress the earth to that size. The surface of a sphere of this diameter is equal to the total area of all of today's continents. At that time, the crust of the earth was still in a hot and liquid state; upon solidification, a relatively uniform crust resulted. The total surface of the earth was covered by a shell whose area was the same as that of all the continents together as we know them today.

Eons came and went. Slowly, extremely slowly, the force of gravity decreased, and the pressure on the interior of the earth diminished. The earth expanded, and the uppermost layers of the crust, brittle enough, broke up into several pieces. Today, these pieces are the continents. The next lower layer, consisting of basalt, was viscous enough so that it did not burst; it merely expanded like the skin of an inflated balloon. In doing so, it became thinner, and it finally developed into the earth's crust as it exists today. This layer of basalt still extends over the entire earth, forming the basis of the ocean floor and also the basis of the continental crustal blocks.

The granite layers of the continental blocks did not expand—

they are still in their original state, and they float as relatively thick slabs, like icebergs, on the plastic substance beneath.

The proponents of this attractive hypothesis include, besides Pascual Jordan, a number of American geophysicists and ocean-ographers. They have pointed out numerous other geological phenomena whose explanation had met with difficulties in the past. For example, they interpreted a worldwide system of fissures in the crust, which has been observed on the ocean floor as well as on the continents, as a direct consequence of the earth's expansion. Also, they were able to establish an entirely new and surprising hypothesis regarding volcanic action, and particularly on the for-mation of mountain ranges.

It is a widespread belief that the mountain ranges were formed like the wrinkles on a shrinking apple. When the interior of an ap-ple shrinks because of loss of water, the skin begins to shrivel, being too large. There is no doubt that the formation of wrinkles has taken place in the uppermost layers of the earth's crust, and that the designation "fold mountains" is still justified.

Obviously there must be horizontal squeezing forces on the sur-face of the continental crustal blocks. The existence of such oroge-netic forces can be explained on the basis of the expansion of the earth. A continental block which was a part of the small proto-earth had a much more definite curvature than now; with the ex-pansion of the earth, it became flatter and flatter. Visualizing this picture, one will readily understand that horizontal squeezing forces must develop on the surface of a crustal block. Cut out about one third of a rubber ball, and try to flatten this cap-shaped segment; it will develop wrinkles.

Those who support the expansion hypothesis are fully aware of the fact that their theory is not yet fully proven. However, their ideas are gaining more and more credibility. The expansion theory is an excellent example of the fact that scientific concepts change, that our knowledge progresses, and that we come closer and closer to scientific truth.

4 ✳ Age of Our Planet

In the discussion of the evolution of our planet, of the formation of the terrestrial sphere, and of the expansion of the earth, it was noted repeatedly that these events must be projected into enormous periods of time—hundreds of millions, even billions of years. Time periods of such duration are entirely beyond our imagination. A millennium can be, comprehended, but our imagination begins to fail in the attempt to review the record of man's history, which spans perhaps ten thousand years. The age of the human race, estimated at about one to two million years, escapes comprehension; this is all the more true of man's ability to comprehend the earth's age, which is several thousand times older.

How such enormous periods can be measured will be questioned by everyone who hears these figures for the first time. Are such measurements reliable? After all, a dependable determination of the age of the earth is highly important, because, without an accurate estimate, reasonable statements could not be made about the course of events that culminated in the shape of the earth as it exists today.

Without careful inquiry, it is impossible to tell the age of a piece of rock picked at random. Such a problem, insoluble for most, is considerable even for the expert. Ask a scientist how old a piece of rock is, and he will immediately reply: "How should the age of the rock be defined? After all, the specimen consists of atoms and molecules, the smallest building blocks of matter. They are indestructible by chemical and geological forces, and therefore must be as old as the universe."

But this is not how the age of a rock, or of the crust of the earth, is normally defined. To know the age of a piece of rock, it is necessary to know when it began to exist as the rock seen at present. In the very early phases of the history of the earth, the entire crust was hot and liquid; later, it solidified—and this was when the rock came into existence. However, this may not always be a satisfactory way to define the age of a piece of rock. Actually, there are rocks just a few days old, such as a piece of lava from a recent volcanic eruption which may even still be warm when it is picked up. To determine the age of a piece of mineral is difficult because, in contrast to living beings, the mineral may not change its appearance over millions of years. Admittedly, changes in rocks are influenced by time, during which erosion by wind and weather occur, or the accumulation of layers of sediments, or even the remelting in hot regions within the crust. Ever since the science of geology began, all these processes have been exploited in order to assess the age of rocks.

It has long been known that these changes take place extremely slowly, and that the history of the earth encompasses a time span of many, many millions of years. However, the forces which bring about such changes have not remained constant throughout the eons. The variations observed in some rocks, therefore, represent a totally inaccurate reckoning. It was left to modern physics to present to the geologists a clock that meets ideally the three requirements demanded of a geological timepiece. First, the clock must move extremely slowly; second, it must be highly accurate;

and third, it must be unaffected by external influences. Only an atomic clock—or, better, a nuclear clock—can fulfill these requirements.

As remarked before, a piece of lava that has cooled and hardened quite recently cannot be distinguished at first glance from a piece that solidified millions of years ago. However, if it contains minute, almost immeasurably small quantities of certain chemical elements, its age can be determined. Such minute quantities of matter by which time can be measured belong to the family of the so-called radioactive elements.

Atoms, the building blocks of the elements, are known to be extremely stable. At least this is true of the atoms of most of the elements found in nature. Enormous energies are needed to destroy an atomic nucleus, which is responsible for the structure of an atom. Disregarding for the moment the negligible proportion of atoms in the air and the uppermost layers of the crust that are being destroyed continuously by the energy of cosmic rays, it may be stated that practically all the atoms present when the earth came into existence are still here now. Many of them have certainly participated in a large number of chemical reactions, but they did not lose their identities in these reactions. The only atoms with a limited lifetime are the atoms of the radioactive elements.

Consider, for example, an atom of the best known of these elements, radium. Each radium atom, at some time during its life, emits a small particle, which is shot off from the parent atom, or rather from its nucleus, at high speed. The nucleus disintegrates into two parts, and with the loss of a small fragment the atom is transformed into another atom. In this process it loses its identity as a radium atom. A piece of radium contains untold billions of atoms; each second, numerous decay processes occur, and a large number of small atomic fragments are emitted at high speed in all directions. Radium continuously emits rays; it is "radioactive."

If one atom were taken at random from a bit of radium, it would be impossible to predict exactly when this particular atom would decay, just as it is impossible at the time of birth to predict the age a human being will achieve. The insurance statistician, in this regard, can only tell the average life expectancy of the baby; which of course is only a statistical figure; but in view of the large number of babies this figure is sufficiently accurate for the computation of a life insurance premium. In a similar manner, a physicist cannot accurately predict the lifetime of an individual radium atom. However, he can tell with considerable accuracy how long it will be before half of all the atoms in the bit of radium will have decayed: 1620 years.

The certainty with which this statement can be made reflects a law of nature that provides great accuracy where large numbers of atoms are concerned, and such numbers are manifest in the case of even tiny bits of matter. Another 1620 years later, or after 3240 years, only one fourth of all the radium atoms will remain. After still another 1620 years, one eighth will be left, and so on.

Clearly, the law of radioactive decay can serve as a clock. If you have a piece of radium, you only need determine its mass and you know at once how much time has elapsed, provided you knew the quantity of radium to begin with. If a geologist wished to determine the age of a piece of ore containing radium, he would be able to establish this by means of the radium clock only if a cooperative scientist several thousand years ago had left a note bearing the date and amount of radium, in milligrams, present in the ore at that time.

Even the best clock is of no use if it cannot be read. Fortunately, the radium clock, like all the clocks in daily use, has two hands, and physicists, chemists, and geologists have learned how to read and interpret their positions.

As stated earlier, a radium atom changes into an atom of another element in the process of radioactive decay. This new atom is again radioactive and decays into an atom of the third generation. This process continues in a chainlike fashion until finally the end product is reached, the element lead, which does not decay further. If at some time long ago a minute quantity of radium had been mixed with a mass of liquid minerals which then solidified into hard ore, that time may be taken as the starting point at which the radium clock began to run. Gradually, imperceptibly, the radium turned to lead. At first the lead did not exist but in the course of time the radium content diminished and the lead content grew. When a geologist finds the piece of ore today he need only determine the ratio of the existing amounts of radium and lead; a simple computation will then produce the age of the ore.

Of course, the liquid mass of minerals might have originally contained some lead along with the radium when the mass solidified. If that lead were now weighed together with the lead resulting from the radioactive decay of the radium, a serious error would occur. It is one of the happiest circumstances for geological research that the original lead can be distinguished from the lead that was produced from the decay of radium. Radium lead differs in atomic mass of 207 units, whereas radium lead has an atomic mass of 206 units. This fortunate situation permits a separation of the two kinds of lead, so that the radium clock can be read unmistakably.

As compared to the long periods of geological history, the radium clock runs quite fast. In round numbers, after about 5000 years only one eighth of the original amount of radium is left, and after 10,000 years, only about one and a half percent remains.

It is evident that the radium clock is not capable of reaching far back into the past history of the earth. Obligingly, however, nature has put at our disposal not one radioactive clock but a whole series of them, which fortunately run at different speeds. The radium clock is nevertheless among the fastest. The uranium clock, for example, is a good deal slower. Uranium also is subject to radioactive decay, although at so slow a pace that a given amount of uranium requires four and a half billion years before it is reduced to half of the original amount.

A large number of age determinations of various mineral samples have been made since physicists and chemists put this excellent time-measuring device into the hands of geologists. It was found that the oldest minerals and rocks must be about three billion years old. This, however, is the age of the rocks in the sense mentioned earlier; the radium clock began to run at the time when the radium mixed with fused minerals that subsequently solidified into hard rock. The oldest rocks found today cannot have been liquid for the last three billion years. The material itself may be much older, because it may have gone several times through the process of melting and resolidification before the three-billion-year count began. On the other hand, the oldest rocks cannot be older than the earth.

It may be stated with certainty, therefore, that the solidified earth is at least three billion years old—probably older; it must be assumed that many of the geological processes, such as the formation of mountain ranges and volcanic action, have caused remelting and resolidification quite frequently during the earth's long history.

Age determination by means of the radioactive clock is accurate and reliable because the quantities of chemical elements in a rock sample can be measured precisely with the tools of modern physics and chemistry. The rate at which the atomic clock runs is entirely independent of external events, as long as the rock sample is not heated to a temperature at which it melts; if it does, the lead might flow away rather than remain within the rock material. There is no physical or chemical process in the earth's crust that would change the rate of the clock, regardless of where the rock has been resting during the millions and billions of years of its existence—at the ocean bottom, inside a mountain, on the surface, or elsewhere.

Accurate though the radioactive clock may be, it measures only the age of the rocks but not the total age of the earth. Fortunately, astronomy, physics, and chemistry have provided several other means of age reckoning. These, however, are not equally accurate.

The distance between the moon and the earth increases slowly in the course of time. What causes this gradual increase? The lunar distance, even in the course of a century, increases so imperceptibly that it cannot be measured, yet the laws of celestial mechanics permit no other conclusion. This effect of recession is caused by the tides. The moon generates two huge tidal waves in the oceans, one of which is always directed toward the moon, the other toward the opposite side of the earth. The earth rotates under these two tidal mountains like a wheel between two brake shoes, and their braking action slows down the earth more and more.

Admittedly, the deceleration of the earth's rotation is minute. In fact, the average length of the day increases by four ten-thousandths of a second per century. The amount is small, but it adds up to several seconds per century; each century is about fourteen seconds longer than the previous one. Astronomical observations have made these calculations possible. Events such as eclipses of the sun or of the moons of Jupiter, which can be foretold decades before they occur, always take place a few seconds earlier than the time predicted. The differences between the calculated and the observed times reflect the retarding effect of the tidal waves; the rotating earth represents a clock that runs a trifle slow.

Earth and moon together form a closed mechanical system. According to the laws of mechanics, the tidal wave exerts the same force on the moon that the moon exerts on the tidal wave. The decrease of the earth's rotational velocity, therefore, will be compensated by an increase of the moon's energy in its orbit of the earth. This minute but continuous transfer of energy from the earth to the moon causes the moon to move away gradually from the earth at a calculable rate.

The moon is presently 240,000 miles distant from the earth. Assuming that the two bodies came into existence as a binary planet at about the same time and in close proximity to each other, it can be computed how much time the moon required to obtain its present distance. The result of this computation is four and one half billion years, a figure that coincides approximately with the age of the oldest rocks, which of course are younger than the earth itself. It should be noted, though, that the accuracy of this method compares less favorably with that of the radioactive clock, because the amount of tidal friction during earlier phases of the earth's evolution is difficult to estimate.

Physics provides yet another method of determining age; this is not applicable to the earth, although it serves to calculate the age of meteorites. Most meteorites are members of the solar system.

They are probably fragments of small planetoids that travel in large numbers around the sun between the orbits of Mars and Jupiter. In the course of billions of years, some of these planetoids undoubtedly collided with each other and broke up into many pieces which have continued to orbit the sun. Occasionally, they strike the earth. The smaller ones burn up in the atmosphere; the larger ones reach the earth's surface. Radioactive elements within these meteorites permit age determination by means of the radioactive clock just as with earthborn rocks. Thus it is known that the oldest meteorites are 4.6 billion years old. Assuming that the earth and the small planetoids originated at about the same time, this figure can be considered as indicative for the age of the earth, or rather for its minimum age.

Surprisingly, there are even methods to determine the age of the atmosphere and of the oceans. Again, it is a fortunate circumstance that the radioactive clock can tell the age of the atmosphere. Almost all the natural radioactive elements are heavy elements; uranium, for example, is the heaviest. The end products of these heavy radioactive elements are the kinds of lead which differ in their atomic masses. There is also a lighter natural radioactive element—a particular isotope of potassium which, when it decays, is changed into argon.

Argon is a noble gas which does not react with any other element. It escapes into the atmosphere and remains there as a gas. The argon generated by radioactive decay of potassium can be distinguished from the original argon in our atmosphere by its atomic mass, which amounts to about one percent of the volume of the atmospheric gas mixture. The total amount of potassium, the parent substance of argon, which exists in the earth's crust can be estimated. By means of this figure, it is possible to compute the time required to generate the argon isotope found in our present atmosphere. As potassium is a relatively rare element, the time must be considerable. The figure arrived at is 4.6 billion years.

The age of the oceans can be estimated in a similar manner. Just as argon accumulated in the atmosphere as a sort of waste product during the long history of the earth, salt accumulated in the oceans. The salt in the sea is indeed a kind of waste product, since it is being washed out of the land by the streams and rivers and carried into the ocean. It remains there because the water enters the second half of its cycle in a pure, distilled form. When sea water evaporates under the heat of the sun, pure water vapor rises; the salt is the residue. The raindrops falling on the rocks form brooks, rivers, and streams which flow across the country, dissolve new salt in the soil, and carry it to the ocean. This process must have started when the oceans began to form, and it has been going on ever since. The amount of salt being carried into the oceans every year can be determined; also, the total amount of salt contained in the sea can be measured. These two figures permit an estimation of the length of time this process must have been going on: several billion years.

The salt method does not provide the same high accuracy in determining the age obtained by other methods. The reason is obvious. It is not altogether correct to say that all the salt that was ever carried into the sea remained there. It so happened off and on during geological evolution that shallow basins were cut off from the oceans. As they dried up gradually, the salt was precipitated; many of the large salt deposits were formed in this manner. It may happen that such a deposit will be washed out again, and that the salt will go through the cycle of precipitation and solution more than once. Another factor of uncertainty in the calculation is the amount of salt flowing into the ocean every year. It is not known how this annual flow may have changed during geological history.

Considering these and other uncertainties in the age determinations, it is surprising that all the various methods provide approximately the same result of four to five billion years for the age of the earth. The age of the oldest rocks turned out to be somewhat less, as expected.

It should be noted that the time span of four and a half billion years does not yet cover the full age of our planet from its very origin. Calculations based on the distance of the moon assume that earth and moon already existed when the moon began to move away from the earth as a consequence of the tidal friction. The age determination of the meteorites assumes that the meteorites had already been formed when their four and a half billion years of existence began; also, the crust of the earth must have been in existence when the decaying potassium began to release the argon into the atmosphere. As mentioned previously, the formation of the earth from the material of the primordial atmosphere of the sun may have required only a hundred million years. This, however, is merely a rough estimate. It is not known how long this phase of the formation of the earth lasted, nor is it known how much time the earth required to heat up to temperatures that permitted the separation and stratification of the core, the mantle, and the crust.

The rates at which such processes occur are difficult to estimate, and a reliable method of measuring these rates has not yet been found. By and large, it may be assumed that these processes during the early part of the earth's history required another billion years. The total age of our earth, then, is five to six billion years.

It is well to conclude the discussion of the age of the earth with a hint about the age of the sun, the stars, the Milky Way, and the entire universe. Of course, physicists and astronomers cannot state the age of the sun or of the universe with absolute accuracy, yet modern science can make estimates about the age of the sun and the universe which are surprisingly reliable.

The age of the sun or of a star can be estimated from the rate at which energy is radiated. According to the laws of physics, a star must exist during its entire lifetime on the store of energy it possessed at its birth in the form of hydrogen. During the lifetime of a star, the hydrogen changes into helium, and into heavier elements, by nuclear transmutation; energy is liberated in these processes. The rate at which a star emits energy can be measured with great accuracy, and its total lifetime can be estimated on that basis.

Lifetimes of stars turn out to be several billions of years. The Milky Way and the other galaxies in the universe consist of individual stars; the age of a galaxy, therefore, must also be figured in billions of years. An astonishing result of astrophysical research is the discovery that the universe is not ten or a hundred times as old as our planet, but only about one and a half or two times as old. In fact, our little earth has been around during a very considerable part of the lifetime of the universe.

5 ❊ Origin of the Atmosphere and of the Oceans

Earth and moon, it is now widely assumed, came into existence at the same time as a kind of binary planetary system. Provided that this is true, it may be concluded that essentially the same chemical elements and compounds were available for the structural makeup of these two celestial bodies. The materials must also have included whatever substances compose the earth's atmosphere and oceans. Following this simple logic, it might be argued that the moon, too, should have oceans and an atmosphere.

A cursory look at the lunar surface with binoculars, or even with the naked eye, reveals that the moon has neither oceans nor air. There is no atmospheric haze that impedes the view on the lunar disk; the moon's mountains, craters, and dark shadows show precisely defined contours. Also, the observer can see no indication that water is present on the moon. If open bodies of water even remotely comparable to an ocean or a large lake existed on the lunar surface, sunlight would occasionally be reflected by them as the light of a lamp is reflected by a sphere of glass. Such reflected sunlight would offer a brilliant display; in fact, it would far exceed the luminosity of the full moon. The absence of atmospheric haze and of brilliant highlights on the lunar surface proves that the moon is without air or oceans. Why is it that the earth has oceans and an atmosphere and the moon does not?

If the moon were suddenly given an ocean or an atmosphere, it would lose it after a short time. This can be shown by a simple imaginary experiment. Suppose a bathtub filled with water were put on the lunar surface while at the same time a container with compressed nitrogen was being vented beside it. The bathtub is needed only so as to prevent the water from seeping into the surface of the moon.

Under the intense heat of the sun, the water in the bathtub would soon begin to boil and evaporate. The particles of water vapor would mix with the molecules of the nitrogen gas, and the mixture would form a highly rarefied lunar atmosphere. After a short time, the ocean, represented by the water in the bathtub, would evaporate completely.

The lifetime of the thin lunar atmosphere, consisting of the molecules of water and nitrogen, is also limited. It is well known that the molecules of gases and vapors are in continuous motion; they move in all directions, and they frequently collide. The higher the temperature of the gas or the vapor, the higher the average velocity of the molecules. Some attain velocities of several miles per second.

Consider one of these molecules, assuming that it moves at a given instant straight upward at a velocity of half a mile per second. This particle will follow a vertical trajectory; it will gradually slow down under the gravitational attraction of the moon, and finally begin to fall back toward the lunar surface. Another molecule whose initial vertical velocity is two miles per second will also be slowed down by the lunar attraction, but the gravitational force is not sufficient to pull the molecule back toward the moon. The molecule will escape from the gravitational influence of the moon, and it will continue to move out into the universe, escaping in the same way that the third stage of a space rocket escapes the gravitational sphere of influence of the earth and goes into orbit around the sun.

On the moon, the escape velocity is 1.5 miles per second. Every particle whose upward speed exceeds this escape velocity leaves the moon forever. Daytime temperatures on the moon are sufficiently high to cause molecules of water and nitrogen to escape the lunar gravitational field after a short time.

The situation is different on the earth. The earth is considerably larger than the moon, and the force of gravity at its surface is approximately six times greater than at the moon's. The escape velocity that enables a gas or vapor particle to leave the earth's gravitational sphere of influence from the highest regions of the atmosphere is more than seven miles per second. Only the lightest atoms, hydrogen and helium, occasionally reach such high velocities.

In a volume of gas, all the atoms and molecules have the same average kinetic energy which is determined only by the temperature of the gas. For this reason, a heavy particle in the volume of gas, such as oxygen, has a much lower velocity than a light particle in the same volume, such as hydrogen or helium. Hydrogen atoms are sixteen times, and helium atoms four times, lighter than oxygen atoms. Fortunately the oxygen molecules in our atmosphere are too slow to escape from the gravitational field of the earth. Hydrogen and helium escape at a slow rate; as a result, our atmosphere contains only small quantities of hydrogen and helium, although the two gases constitute 99 percent of the total mass of the universe.

The earth holds its atmosphere in a firm grip, and the oceans, too, are retained by the same strong gravitational force. It is true that the sun vaporizes a huge amount of water from the oceans every day, and that this water vapor mixes with the atmosphere. Each water molecule consists of an oxygen atom and two hydrogen atoms. It is eighteen times heavier than a hydrogen atom, too heavy to escape the gravitational pull of the earth. Sooner or later, almost every water molecule that finds its way into the atmosphere will attach itself to a snow crystal or a water droplet, and thus will return to the earth and into the ocean.

The statement that the atmosphere contains only small amounts of hydrogen refers only to pure hydrogen gas. As soon as the hydrogen atoms are chemically bound to heavier atoms, as in water molecules, they are parts of heavier molecular masses and therefore cannot escape. Practically all the hydrogen in the sea and in the lower atmosphere is a constituent of heavier chemical compounds.

This brief excursion into the chemistry and physics of gases was necessary to show how the earth acquired oceans and an atmosphere during the early phases of its evolution—also why the present atmosphere is the second, perhaps even the third or fourth, developed during the earth's existence. It is important to remember here that the earth originated from a gigantic vortex of cosmic material composed of gases, dust particles, and larger accretions of matter. When the mass of the young earth was large enough to prevent gases from escaping from its field of gravity, those gases began to form a proto-atmosphere around the earth which did not condense at the relatively low surface temperature of about 60° F. Later, the earth heated up under the continuing impact of accumulating material, by gravitational contraction, and by the energy released from radioactive elements. Its temperature rose to the melting point of the materials even at the surface, and the proto-atmosphere of the earth, too, assumed a high temperature.

The average velocity of particles in a gas increases with rising temperature; therefore, when the proto-atmosphere became hotter and hotter, the gravitational force of the earth could no longer prevent even the heavier molecules from escaping. This is how the earth must have lost its first atmosphere some hundred million years after its birth. That the earth lost this proto-atmosphere is shown by an incontrovertible argument, so cogent, in fact, that it also proves that at least once during its early history the earth's surface must have been extremely hot.

The proof is based on the abundance of the noble gases in the atmosphere. Consider, for example, neon, a gas commonly used in lighting. The atoms of neon do not combine chemically with any other element. They are heavy enough to be retained by the gravitational field of the earth. Astrophysicists have found that neon is a relatively abundant element in the universe. In the earth's atmosphere, it is extremely rare, which is why neon was discovered quite late. Since neon gas does not react chemically with other substances, it should be expected that neon would occur in the atmosphere with its original great abundance. In view of its rarity, the conclusion is unavoidable that practically all the neon escaped into the universe at the time when the proto-atmosphere of the earth was lost.

However, it could escape only at high gas temperatures, which proves what has just been stated—that during its early history the earth's surface must have been hot indeed. This chain of argument illustrates the close cooperation between different scientific disciplines that makes it possible to find answers to questions that otherwise might remain forever unanswered.

When the crust of the earth began to cool and solidify, a new atmosphere formed, probably within a relatively short time. Volcanic action must have been severe at that time. Although large areas on the earth's surface were covered with cold cinders, there remained extensive fields of red-hot lava, along with other materials erupting from volcanoes, that contained large quantities of vapors of various substances. The vapors and gases emanating from the liquid matter through vents and fumaroles were composed of various elements.

Originally the earth consisted of a typical mixture of universal matter. The elements in this mixture were available for the gases and vapors that were expelled in volcanic eruptions. The most abundant elements, in descending order, were hydrogen, helium, carbon, nitrogen, and oxygen. Hydrogen and helium were the chief elements. As an ingredient of an atmosphere and of the oceans, helium is negligible; it does not react chemically with other elements, and its individual atoms are so light that they eventually escape from the atmosphere into space, which is why helium is a rare element on earth, although it is the second most abundant material in the universe. Hydrogen, however, is chemically very active. At an early time, it formed compounds with the other elements that were abundant on earth, especially carbon, nitrogen, and oxygen.

The chemical structure of the earth's atmosphere reflects in a very direct sense the elements that were available, and their mutual reactions. Hydrogen was overly abundant during the early phases of the earth's history; the first compounds it formed were methane (CH_4), ammonia (NH_3), and water (H_2O). There was enough hydrogen in the material of the original earth to bind all the available atoms of carbon, nitrogen, and oxygen in the form of methane, ammonia, and water. Large stores of these compounds developed in the interior of the earth. They were available to form the second atmosphere when the surface of the young earth began to cool and solidify.

When the earth's surface had cooled down sufficiently, the water vapor condensed in the form of water and began to fill the ocean basins. Ammonia and methane remained in the atmosphere as gases. This was the second atmosphere of the earth; it must have covered our planet about four billion years ago. Presumably, the total amount of water in the proto-oceans was much less than today; the crust of the earth has been producing new water ever since that time. The exhalation products of volcanoes still contain more than 90 percent of water vapor that rises to the surface from deeper layers in the earth's crust. In Mexico, there is a small volcano that produces an unusual quantity of water vapor. At the rate of its present yield, this volcano alone would have been able in the course of four billion years to provide all the water in the oceans of the world. This example supports the belief that the oceans did not come into existence in their present size. They must have grown throughout the history of our planet, and they are still growing today.

Shortly after the earth's crust solidified, the oceans began to form, at first with relatively little water. The atmosphere consisted then mainly of methane and ammonia. As ammonia dissolves readily in water, it must be assumed that the early oceans contained large quantities of ammonia in solution.

It is interesting to note that the atmospheres of the large outer planets, Jupiter, Saturn, Uranus, and Neptune, consist almost totally of methane and ammonia. Obviously, these planets have not developed atmospheres beyond the methane-and-ammonia phase. On the earth, powerful processes must have been at work which led to a fundamental change of its atmosphere. Today, the earth's atmosphere contains no methane or ammonia; it consists almost wholly of nitrogen and oxygen.

Yet, in former times, methane and ammonia were the principal constituents of the earth's atmosphere. Some force must have been active that was capable of breaking up the chemical bonds of these gases. This force was provided by the sun. Solar radiation contains about 7 percent of ultraviolet light, whose energy is so great that it can produce a number of chemical reactions, among them a severe sunburn. Ammonia and methane were exposed to this ultraviolet radiation for millions of years. Its quantum energy is sufficient to decompose the molecules of ammonia and methane into their original atoms. Eventually, ammonia was split into hydrogen and nitrogen, and methane into hydrogen and carbon. The water vapor in the atmosphere also was split into its components, hydrogen and oxygen. Hydrogen atoms and molecules, as previously described, move too fast even at moderate temperatures to be retained by the gravitational field of the earth. They escaped into space, and only nitrogen, oxygen, and carbon remained in the atmosphere.

Gaseous nitrogen is chemically inactive toward many other elements and compounds, including oxygen and carbon. However, oxygen and carbon have a great mutual affinity. Thus nitrogen remained in the atmosphere as a gas, but carbon and oxygen reacted to form carbon dioxide, a gas well known from the bubbles it forms in soda water.

In the course of millions of years, the atmosphere of the earth changed, under the influence of the sun's radiation, from a combination of ammonia, methane, and water vapor to nitrogen and carbon dioxide. The question may be asked why the same processes did not occur also on Jupiter, Saturn, Uranus, and Neptune. Those planets retained their early atmospheres simply because their gravitational fields are strong enough to retain even their hydrogen. Besides, the intensity of solar radiation is much lower on these planets than on earth because they are farther away from the sun. The few hydrogen atoms that occasionally were split from ammonia and methane molecules soon found other partners with which they formed new ammonia and methane molecules. For those reasons, the early ammonia and methane atmospheres of the large planets have been stable from their beginning to the present day.

The nitrogen and carbon dioxide atmosphere was the third in the history of our planet. Obviously, that was not yet the atmosphere the earth has today. The sister planet Venus still has its nitrogen and carbon dioxide atmosphere. It seems to have stabilized in this phase of atmospheric evolution. Venus, too, presumably once had an atmosphere of methane and ammonia, but the transformation of those atmospheric constituents into nitrogen and carbon dioxide must have occurred even faster on Venus than on earth. Venus, being about the same size as the earth, must also have lost its hydrogen relatively soon. The loss rate must have been even greater because Venus is closer to the sun, and the intensity of the solar radiation is more than twice that on earth.

A brief excursion into the history of the atmospheres of our neighbor planets may help us to understand the development of the earth's atmosphere. It is assumed with reasonable certainty that all the terrestrial planets came into existence at about the same time. Principally, they differ in only two of their properties: size, and distance from the sun. As noted before, these two factors exercise a decisive influence on the development and the structure of a planetary atmosphere.

A small planet, or a small moon, is incapable of retaining any atmosphere at all. A planet the size of the earth or Venus cannot retain the most abundant and chemically important gas, hydrogen, unless it is chemically bound to other elements. Carbon, nitrogen, and oxygen can be retained. A very large planet will not lose even its hydrogen. Planets that are as close to the sun as the earth and Venus will experience a strong interaction of the solar ultraviolet radiation with the constituents of their atmosphere, whereas the atmospheres of planets that orbit the sun at great distances will not be significantly affected by solar radiation. These simple relationships will become more evident in a planet-by-planet review of our solar system.

The planet nearest to the sun, Mercury, is only a little larger than the moon. As explained earlier, the moon is too small to retain an atmosphere or an ocean. The same is true of Mercury. Venus has an atmosphere which consists mainly of carbon dioxide, probably mixed with some nitrogen and argon, and a trace of water vapor. Next in order is the earth, whose atmosphere has already been discussed. Mars is the fourth planet, to be discussed presently. The outer planets Jupiter, Saturn, Uranus, and Neptune still have their old atmospheres consisting of a mixture of ammonia and methane. The reason they retain this atmosphere is due to their size and their distance from the sun. They are large enough to retain the light hydrogen gas, and far enough from the sun to avoid the decomposition of ammonia and methane. These gases will probably be stable on those planets for an indefinite time. The atmospheres of Jupiter, Saturn, Uranus, and Neptune have not experienced the metamorphosis that took place in the atmospheres of Venus and the earth.

As to Mars, next to the smallest of the planets, it is larger than Mercury, but its mass is only one tenth that of Venus or the earth. The force of gravity at its surface is much less than that of the earth or Venus; as a consequence, Mars quickly lost its hydrogen at every phase of its development. The history of its atmosphere differs substantially from that of other planetary atmospheres. After the earth had lost its proto-atmosphere, water vapor erupted from its interior and led to the formation of the oceans. On Mars, the production of water vapor presumably has always been much less than on earth; Mars' solar radiation decomposed the water vapor into hydrogen and oxygen at about the same rate as it was produced. The hydrogen gas escaped quickly into space, and the remaining oxygen was consumed in the oxidation of the Martian rocks.

This is the reason for the reddish color of Mars; very probably, this color is caused by an abundance of iron oxide on its surface. As behooves an ancient war god, its armor has accumulated some rust in the course of billions of years. Only a small part of the oxygen was consumed in the formation of carbon dioxide, a conspicuous process in the formation of the earth's third atmosphere. The bulk of the present Martian atmosphere consists of nitrogen, possibly mixed with some argon.

The color of a planet is the result of its distance from the sun and the nature of its atmosphere. The dense veil of clouds that shrouds Venus gives this planet a brilliant white glare. On Mars, whose atmosphere is so thin, color is determined by the reddish desert over most of its surface. The four large planets—Jupiter, Saturn, Uranus, and Neptune—are also enveloped in heavy cloud layers. In each, color is determined by the chemical composition of their atmospheres. Ammonia and methane are absorptive, causing the greenish hue reflected by these planets. The more distant they are, the more definite is that greenish tinge as seen from earth.

On our own planet, the atmosphere scatters blue light in all directions, and that scattering effect accentuates the color. The sky, viewed from the earth's surface, appears blue, a color that is radiated out into space, which is why the astronauts see it as if the earth were enveloped in a cerulean shroud. In areas where there are no clouds, land surfaces and outlines are plainly seen. Wide stretches of ocean are often distinctly visible, and their color also adds to the quality of blueness, unique among the planets.

6 ⁕ Origin of Life

How life came into being has occupied the minds of men since time immemorial. The phenomenon of life on earth was touched upon earlier, in connection with the origin and evolution of the atmosphere during the earth's geological history. Scientists are in general agreement that it was not until the third atmosphere, about three billion years ago, that life was possible. If life had not developed then, this atmosphere would still persist.

The chemical activity of the green plants, through the process of photosynthesis, separated the carbon dioxide of the air into its components carbon and oxygen. The plants used the carbon to build up their tissue structures; the oxygen was given off as a waste product. This is how the present atmosphere was formed. It consists of 80 percent nitrogen from the original third atmosphere and 20 percent oxygen from the chemical factories of the green plants.

The existence of free oxygen in the atmosphere of our planet is not an inherently stable circumstance. If plants were suddenly to cease their chemical activity, most of the oxygen in the earth's atmosphere would disappear after as short a span as three thousand years. The oxygen would be absorbed through the oxidation of minerals in the uppermost layers of the earth's crust, and by the decay of organic matter. It would not be correct to assume that the green plants produced all the oxygen presently existing in the air at one time long ago; actually, they replenish the enormous oxygen losses that occur through oxidation and organic decay. The oxygen in our atmosphere owes its existence to a delicate state of dynamic equilibrium which the green plants maintain by means of their highly efficient chemical activities.

Oxygen is chemically so active that it does not occur in a free and gaseous state on celestial bodies under normal circumstances; it forms chemical compounds with other elements soon after it has been produced. The earth is the only planet in the solar system whose atmosphere permanently contains substantial amounts of free oxygen. Consequently, in all probability, no other planet in the solar system carries as large a number of living creatures as does the earth.

The paramount importance of life on earth for the chemistry of the earth's atmosphere has been recognized in its full significance only during the last few decades. Indeed, the evolution of our planet, at least in its later phases, cannot be understood unless careful consideration is given to life and its evolution.

Man himself is only one species of living beings on earth. This may be why it took a long time before the question of the origin of life could be studied and discussed without prejudice. Even during ancient times, men with searching minds were convinced that the higher forms of life could not have come into existence spontaneously, but must have descended from parents. When this logic was applied in retrospect along the chain of past generations, a point was reached at one time or another when men had to resort to a mythical explanation of the origin of man and the higher animals.

For a long time it was believed that lower forms of life came into existence spontaneously; worms were thought to originate in mud, maggots in decaying meat, and mice in moist hay. It is common knowledge today that such conclusions arose from unscientific, superficial observations. However, only a few hundred years ago, leading scientists still believed in spontaneous generation, among them the English physician William Harvey (1578–1657), the discoverer of the mechanism of the circulation of the blood, and Isaac Newton.

Two Italian scientists, the physician Francesco Redi (1626–1697) and the biologist Lazaro Spallanzani (1729–1799), demonstrated that no maggots developed in a piece of meat isolated from its surroundings by a simple wire screen. It is well known today that maggots will not develop in meat if flies are prevented from laying their eggs in the meat. Only in the last century, the French physician Louis Pasteur (1822–1895), in a series of brilliant experiments, proved beyond doubt that life cannot develop under sterile conditions.

Pasteur's experiments posed a difficult problem for scientists. Experiments showed clearly that even the most primitive living beings must have ancestors. Following the ancestral line further and further back, one cannot escape the question of how and when the very first living being was created. Before the problem can be attacked with any hope of success, another question must first be studied in great detail. What is the material composition and structure of living organisms? The answer to this question is the objective of biochemistry. The train of thought of leading biochemists who have studied the origin of life will be outlined briefly here.

Living beings, like all other material things in nature, consist of atoms and molecules. Soon after the beginning of modern chemistry it became apparent that all matter found in nature can be divided into two classes, inorganic and organic. These designations indicate the direction of approach to the problem. Inorganic matter consists of molecules, or composite atomic structures, in which only a few atoms have combined to form the molecular compound.

Water, salt, and sulfuric acid are examples of inorganic substances. Organic matter consists of molecules containing as many as tens of thousands of atoms, and even more.

These molecules are highly complex and variable; the number of their possible combinations is almost boundless. This fantastic capacity of combination and variation of organic molecules is owing to the property of carbon atoms to combine with other carbon atoms and to form long chains, rings, and helical structures as basic frames of large molecules. In these molecular systems, carbon atoms combine with hydrogen, oxygen, nitrogen, phosphorus, sulfur, iron, magnesium, and a large number of other so-called trace elements of living matter.

Just a few of the many different molecular structures that form the substrate of a living organism will be mentioned here. Among the most important ingredients of living matter are the proteins. These have extremely complex molecular structures composed of amino acids. Thus far, nineteen different amino acids have been identified. The molecules of most of the proteins have helical structures of different combinations and subunits, each molecular subunit consisting of tens or hundreds of thousands of atoms. Other ingredients of organic matter are fats, starch, and cellulose, built up from long molecular chains.

During the last twenty years, the chemical analysis of life concentrated on the nucleus of the cell, which consists of correspondingly named nucleic acids. The chemical analysis of the genes, the carriers of heredity, was a brilliant achievement of modern biochemistry. Genes are parts of giant molecules in which certain submolecules are arranged in periodic patterns resembling the dots and dashes of Morse code. The giant molecules are long chains; each molecule is structured like a double helix whose halves match each other in specific patterns. The substance of these molecules consists of deoxyribonucleic acid—the well-known DNA. The discovery of the structure of DNA is one of the most significant landmarks in modern science, because it provides an understanding of the growth and propagation of living beings.

When a cell divides, the two matching halves of the double helix separate, and each half gathers from its environment those molecular subunits it needs to rebuild a complete double helix.

The structure of a DNA molecule is highly complex. Each molecule owes its existence to another DNA molecule of identical composition which existed before it came into being. The discovery of the DNA molecule, therefore, does not yet answer the riddle of the origin of life. This question, however, is no longer a chicken-or-egg matter, but rather, where did the first DNA molecule come from? Assuming that one such molecule existed during the remote past, it is now understandable how this molecule, and countless other DNA molecules following it, reproduced exact replicas of the parent molecule, and how the endless variety of present-day life finally came into being.

Proteins, fats, starch, sugar, nucleic acids, and all the other complex organic substances contain as main elements carbon, hydrogen, nitrogen, and oxygen. In addition, they contain trace elements such as phosphorus, sulfur, calcium, and others. The four main elements are those substances which existed in large quantities in the oceans and in the atmosphere billions of years ago.

At that time, the earth already contained large oceans, and also an atmosphere consisting originally of ammonia, methane, and water vapor, and later of nitrogen and carbon dioxide. The transition from an atmosphere consisting of ammonia and methane to one of carbon dioxide and nitrogen did not occur suddenly. It probably took one or two billion years. During this transition, all four gases existed in the atmosphere, and also in the oceans, because they are soluble in water, with the exception of methane. Therefore, all those chemical elements that constitute living matter existed in the oceans early in the earth's history. How did these elements get together? How were they combined and organized into these tremendously complex molecules which form the building blocks of living organisms?

One cannot expect that a DNA molecule would develop in a cup of water containing carbon dioxide, nitrogen, ammonia, and oxygen in solution. In order to cause this stable and inert mixture of solutions to undergo new chemical reactions, a burst of energy is necessary, such as an electrical discharge in the form of a lightning bolt, or energetic radiation from radioactive elements or from cosmic rays. All these sources of energy existed on earth at the time when life is assumed to have originated. Stimulated by this thought, scientists carried out a number of significant laboratory experiments. In 1951, the American biochemist Melvin Calvin exposed a mixture of water, carbon dioxide, and hydrogen to energetic radiation. As a result, he obtained several typical organic substances, among them formic acid, acetic acid, and certain more complex organic acids. A year later, Stanley Miller, a student of Nobel Prize winner Harold Urey, treated a mixture of ammonia, methane, hydrogen, and water with an energetic electrical discharge. After a week's time, the solution contained the two simplest amino acids, glycine and alanine.

Acetic acid and glycine are the two basic materials from which living organisms manufacture the so-called porphyrin ring. Chlorophyll, the green substance in plants, is a porphyrin. It is indeed possible, even likely, that organic molecules were generated in large numbers in the primeval oceans as a result of lightning strokes, cosmic rays, and radioactive radiation from minerals in the earth's crust. At some time eons ago there was probably one molecule among billions of other molecules that was capable of reproducing itself, like the DNA molecule. That was the beginning of life. At that time, the oceans must have contained a rich mixture of organic molecules of many different kinds. Bacteria, which cause the decay of organic matter, did not yet exist. As soon as the first DNA molecule had come into existence, it found sufficient material in its vicinity to reproduce and propagate its kind in abundance. How long this growth continued is unknown, but it is likely that within a relatively short period of time all the organic molecular subunits in the oceans were used up in the formation of higher organic molecules.

At the end of this period, the neophyte life encountered its first crisis. It suffered from famine, because new organic submolecules, the food of the higher organic molecules, developed only very slowly as the result of electrical discharges and energetic radiations interacting with ammonia, methane, and water.

During this phase in the evolution of life, the second great innovation of live molecules must have been introduced—photosynthesis with the aid of chlorophyll. Chlorophyll is capable of synthesizing complex organic molecules by means of sunlight. The incidental formation of the first DNA molecule would have remained without consequence if the formation of the first chlorophyll molecule had not taken place around the same time. Very likely the two basic constituents of chlorophyll, acetic acid and the amino acid glycine, were abundant in the early oceans. They can be generated relatively easily in laboratory experiments, as described.

Assuming that DNA molecules and chlorophyll molecules had come into existence at some time in the remote past, the further development of living organisms can be traced without difficulty. Chlorophyll, with sunlight as a source of energy, is capable of decomposing water and carbon dioxide into their basic components, carbon, hydrogen, and oxygen. Out of these basic ingredients, chlorophyll then synthesizes the numerous building blocks of living organisms, sugar, fat, starch, and, by adding nitrogen and phosphorus, amino acids and nucleic acids. The DNA molecule, on the other hand, is responsible for the unceasing reproduction of living matter.

A visit to a biochemical laboratory will give the visitor a profound impression of the complex and tedious procedures that are necessary to produce artificially even small organic molecules. Is it reasonable to assume that complex organic molecules came into existence accidentally? One is tempted to answer this question with a flat No. However, it should not be overlooked that nature took a tremendously long time to initiate this process. In nature's recipe for the creation of life, time was certainly the most important ingredient. During the long history of the earth, more than two billion years were needed for the generation of the first DNA molecule and the first chlorophyll molecule—more than twice the age of the oldest fossil trilobites.

The work cycle of the chlorophyll molecule is a fascinating study. Green plants, in the course of photosynthesis, split the carbon dioxide which they extract from the air or from water. Carbon is used in the manufacture of the plant structure, while oxygen is released back into the atmosphere or the water. It is not known how many million years it took the plants to consume most of the carbon dioxide of the third atmosphere, and to enrich the oxygen content of the ensuing atmosphere to its present 20 percent. A long time after the first plants had appeared on earth, animals began to develop which breathe the air and use the oxygen to energize the motors of their life processes.

Without life, the development of the earth's atmosphere would have been far different. Indeed, life and its fantastic evolution have made the earth into the planet we inhabit and know today.

The question of the possible existence of life on other planets is inescapable. It is not likely that life exists on other planets of the solar system, with the possible exception of Mars. Mercury and Venus are far too hot, and Jupiter, Saturn, Uranus, Neptune, and Pluto are far too cold. Under their environmental conditions, life cannot evolve. The delicate, highly complex molecules that life depends on are extremely sensitive to temperature. Even at a moderate rise in temperature these organic molecules decompose. At temperatures substantially below the freezing point of water, their reaction rates are slowed down so that life processes virtually cease.

For some time, it had been speculated that Venus might be the abode of a rich abundance of life. Such hopeful expectations, however, have come to an end, at least for the time being, because of a brilliant experiment in connection with the space flight program. In December of 1962, the first successful American instrumented space probe, Mariner II, flew past Venus on a close approach path and took fairly reliable temperature readings. Data were radioed back to earth, revealing a surface temperature on Venus of over 800° F. This high temperature leads to the conclusion that the surface of Venus cannot support life as we know it on earth.

In view of the decisive influence which life exercises upon the chemical composition of a planetary atmosphere, this negative result was to be expected. The Venusian atmosphere does not contain observable amounts of free oxygen. On earth, as pointed out, oxygen is being generated continuously by vegetation. Venus is still surrounded by an atmosphere rich in carbon dioxide. The plants on earth, of course, are responsible for the almost complete removal of carbon dioxide from the terrestrial atmosphere.

On Mars, large areas can be observed whose colors change with the rhythm of the seasons. In the Martian spring and summer, these areas show conspicuous blue-green hues. The green color, and the seasonal cycle, may be caused by primitive forms of vegetation. However, the existence of life forms on Mars would still be a puzzle because the origin and evolution of life would require, at least for a period of several hundred million years, the existence of oceans. It is unlikely that Mars ever had oceans. So it is difficult to understand how life could have developed on Mars. On the other hand, there is a possibility that living organisms reached Mars from outer space. For example, it is not impossible that minute bits of living matter were transported from the earth to Mars by radiation pressure, and that terrestrial life, so to speak, has fertilized that planet.

These problems are equally exciting from scientific and from philosophical viewpoints. The first observational results of the landing of an instrumental spacecraft on the Martian surface, therefore, are anticipated with the keenest interest.

7 ✻ Development of Climates

      The meaning of climate is difficult to define. In essence, the climate represents the sum of all environmental factors, such as temperature, humidity, and wind, which are significant for living organisms. In a more restricted sense, a climate can also be defined for life in the water. Environmental factors in the water include temperature and flow conditions. Even the chemical makeup of the environment of living organisms is to be counted among the climatological factors. Furthermore, the various radiations to which living organisms are exposed must be considered as part of the climate, such as the heat radiation, the visible radiation, and the ultraviolet radiation from the sun—these represent the radiation climate. Energetic radiations from the radioactive elements in the earth's crust, and the cosmic rays that continuously arrive from space and which impinge also upon living beings, contribute to the radiation climate.

      Man, being constantly exposed to climate, has a vivid impression of its overwhelming influence upon his life. During the long history of the earth, the climate on our planet has been subject to many changes. In fact, it is not necessary to reach back very far into the past to find a period during which significant climatological changes occurred. Fossils dating back several hundred million years indicate that substantial climatological changes must have occurred during that period. This fact is borne out by the coal-bearing strata in Antarctica which prove that the sixth continent, presently buried beneath a shield of ice one to two miles thick, must have been covered by green forests millions of years ago. The fact is further attested by the Ice Ages that occurred during the recent past of the earth's history, at a time when man already roamed the planet. In Siberia, the frozen bodies of mammoths were found which were completely preserved, even with their soft parts, in the permafrost of the soil.

There are three basic climatological zones on earth: the tropical zone, the temperate zone, and the arctic zone, all characterized primarily by their temperatures. Most people live in the temperate zone; the tropical and arctic zones imply extremes of high and low temperatures. The differences, of course, are caused by the differences in solar radiation to which these zones are exposed. However, the temperature differences between the tropic and the arctic zones are surprisingly small—on the average, no greater than 100° F., rarely as much as 180° F. Disregarding the extremely low temperatures sometimes observed at the poles during the polar night, the temperature differences over a wide area on the earth's surface are not greater than about 50° F.

As compared to terrestrial temperature variations, the changes in temperature on the moon are remarkable. During the lunar day, under the full impact of solar radiation, the temperature on the moon's surface reaches more than 200° F., but toward the end of the lunar night, which lasts fourteen earth days, it drops to about minus 210° F. Although the average distance of the moon from the sun is equal to the earth's solar distance, the temperature of the moon varies far more than that of the earth's surface. The reason for this difference is the fact that the earth has oceans and an atmosphere, both in constant motion, and by this motion they distribute the solar energy over the entire globe. Besides, the oceans with their enormous quantities of water represent an ideal heat buffer. Covering 71 percent of the earth's surface, they absorb solar energy almost as a sponge absorbs water. The ocean temperature rises but little in this process because of the large heat capacity of water. On the other hand, the oceans release the stored heat energy only slowly during the night and in winter. This is the reason why the temperature of the sea at any given place varies by only a few degrees between day and night and between summer and winter.

The equalizing effect which the ocean exercises on the temperature of its environment is well known from the difference between maritime and continental climate. Coastal regions have moderate summers and relatively warm winters, whereas areas in the interior of large continents, far away from the balancing influence of the ocean, have extremely hot summers and fiercely cold winters.

All living organisms, including man, react with great sensitivity to variations in temperature, so that the oceans on our earth are of utmost importance with respect to the climate and its influence on the development of life.

Such aspects are essential to a study of the climates of past eras. A discussion of paleoclimatology may choose between two avenues of approach. The first would begin with a listing of geological and paleontological arguments that lead to the conclusion that certain changes in climate must have occurred in the past. Based on these arguments, an explanation of the origin and the variations of climate could then be attempted. The second approach would begin with modern theories of the genesis and the evolution of the earth. Changes in the climate could then be deduced step by step as logical consequences of the changes that occurred on the earth's surface.

The second premise is the more attractive; it concerns the hypothesis of the earth's expansion, which has led to a profound revision of our concepts of geological evolution and the development and changes of climate during geological times.

It will be recalled that Paul Dirac developed the hypothesis that the gravitational constant, which governs the mutual attraction of masses, has been subject to a slow but continuous decrease during the history of the universe. Many leading scientists have accepted this hypothesis as a source of new thought and knowledge regarding the evolution of the sun, of the solar system, and particularly of the earth. This leads to the assumption that the diameter of the earth has been slowly increasing, until today it is about twice as large as when the earth came into being.

The size of the earth does not directly influence its temperature and the climate. However, the decrease of the gravitational constant must have had a decisive effect on the climate of the earth, though indirectly, during the early phases of the earth's evolution.

The law of decrease of the gravitational constant must have been valid also for the solar sphere. Billions of years ago, the sun must have been smaller, because the greater gravitational force compressed it more than at present. The processes governing the production of energy in the interior of stars are fairly well known. The emission of radiative energy by a star increases as the pressure and temperature of its interior increases. Provided that the gravitational constant has indeed lessened in the course of time, the sun must have been much brighter some three, four, or five billion years ago. In addition to a higher energy production by the sun, the greater gravitational constant at that time also caused the planets, including the earth, to orbit the sun at shorter distances. This effect led to a further increase in the intensity of solar radiation impinging upon the earth during the earlier phases of its history.

What kind of climate prevailed in the distant past, assuming that the solar radiation was far more intense at that time? In the case of a celestial body like the moon, which has neither ocean nor atmosphere, the question is easy to answer. The daylight temperature on its surface must have been much higher then. The situation was quite different on the earth. The stronger radiation simply caused greater evaporation from the oceans, and the water was suspended in the atmosphere in the form of clouds.

It must be assumed that, during the early phases of the earth's history, layers of cloud covered the planet from pole to pole, so thick that only dim twilight existed on the surface, while much of the sun's radiation was reflected back into space. The reflecting power of clouds becomes evident when the clouds are viewed from an airplane at high altitude; they appear brilliant white.

This thick, enveloping cloud formation prevented the build-up of high temperatures on the earth despite the fact that during those early times the intensity of the solar radiation was considerably higher than now. It shrouded the earth for a long time, probably one to three billion years. Strangely enough, it produced an almost uniform surface temperature over the entire globe, estimated at 50° to 100° F. from pole to pole.

In all likelihood, life originated during the last phase of this climatologically uniform era. At that time, the first chlorophyll molecule must have been generated. Chlorophyll, it will be recalled, is capable of splitting the molecules of carbon dioxide and of water, and of using the products to build up organic substances for the structural elements of plants. Scientists have always marveled at the fact that minute quantities of light are sufficient for photosynthesis. Even in twilight and under water the process functions well.

As the earth became larger and larger, the continental crustal blocks began to separate, leaving large troughs between them. The blocks began to protrude from the oceans and to form huge islands of dry land. Gradually, life began to spread out beyond the sea and to take possession of the land.

Hundreds of millions of years have passed since that period when this strange condition of uniform, humid, tropical climate, with twilight illumination beneath a dense layer of clouds, prevailed on earth. Yet even today jungle plants thrive best in the dim twilight of a tropical environment. They are the nearest modern relatives of the primitive plants of past eons.

During this period of uniformly humid and warm climate, the continents from pole to pole were probably covered by dense, tropical jungle, judging from the existence of huge coal-bearing layers in Antarctica and fossil palm trees in regions like Spitsbergen. The discovery of fossil remnants of tropical growth in areas now belonging to the arctic zones had hitherto mystified paleo-climatologists. Earlier findings led to numerous suppositions about the shifting of the poles in the course of the earth's history which might have changed Antarctica and Spitsbergen into tropical regions.

However, there are arguments against the assumption of a worldwide tropical climate. Along the present equatorial zone, vestiges of a huge ancient glaciation have been found, extending around the globe. Even if the poles had shifted as far as the present equator, traces of this glaciation could be expected at only two diametrically opposed places in the equatorial zone, near the points where the poles were once located.

Pascual Jordan, who strongly supports the expansion hypothesis, suggested an ingenious explanation for the ancient glaciation at the equator. At the time when the earth was shrouded by a dense cloud layer, and when a humid, tropical climate prevailed, heavy rains must have poured down almost incessantly. The force of the thunderstorms and rain squalls in the equatorial zone must have been tremendous. Even now, thunderstorms near the equator are more violent than anywhere else. Places having the heaviest atmospheric turbulence are often the scene of another kind of violent meteorological reaction—hail. Summer thunderstorms are frequently accompanied by hail, and it is not impossible that in times long past the abundant and continuous precipitation of hailstones formed a permanent glaciation, and that this glaciation covered practically the entire area along the equator.

As the gravitational constant slowly diminished, according to Dirac, the size of the sun increased, and the intensity of its radiation subsided. At the same time, the earth increased its distance from the sun because of the diminishing attraction between the two celestial bodies. These two effects, of course, resulted in a decreasing irradiation of the earth by the sun. The lower layers of the atmosphere became cooler and cooler, with the result that the atmosphere could no longer retain those enormous quantities of water vapor which had formed a dense cloud layer for millions of years. The quantity of water pouring down in rains exceeded that which evaporated during the same period, and the water content of the atmosphere decreased still more. The cloud layer became thinner, the days turned brighter, and finally there came a time when at last the sun broke through the clouds. This happened perhaps five hundred million years ago. Ever since, the earth's climate has changed only insignificantly. Life had already spread over most of the earth's surface. A journey by time machine that could take us back four hundred or five hundred million years would show little difference between the climate then and now.

On the day when the first rifts appeared in the dense cloud layer, our planet changed color. While the cloud cover remained dense the earth would have appeared in a brilliant white glare to a viewer from space, because it would have reflected the entire visible spectrum of the sun's light. As the cloud layer became thinner and thinner, the earth assumed a more bluish hue until, as the ocean areas showed through, the earth became transformed into the blue planet seen today through the window of a vehicle in space.

Of course, there were periods on earth during which the mean annual temperature was higher or lower than the long-time average. However, these variations were too insignificant to endanger the existence of life on earth.

Using the fact that the metabolism of mollusks is highly sensitive to water temperature, American scientists have given fascinating proof that the ocean temperature has been subject to only minor variations. As their shells develop, mollusks deposit calcium carbonate, which contains oxygen. Atomic oxygen consists of two kinds of atoms, so-called isotopes, which differ in their atomic masses. When the water temperature varies, the ratio of the rates at which the two oxygen isotopes are deposited in the shells varies accordingly. This correlation makes it possible to determine temperature variations of sea water during past eons, simply by analyzing the calcium carbonate in fossil shells.

One interesting and important chapter in the science of climatology deals with the ice ages—relatively short spans in the history of the earth during which the climate was severe. The designation is, of course, an anthropomorphic qualification; however, when the polar caps were reaching deep into the temperate zones, and when large areas were covered by glaciers, these regions were made barely habitable to man.

At least three definite periods of glaciation are recorded in the geological history of the earth: the first, four hundred to five hundred million years ago; the second, two hundred to two hundred and fifty million years ago; and the last, in the recent past of the earth, a few million years ago. The last period of glaciation ended only about twenty-five thousand years ago. Its imprints are still fresh, and they have provided a fairly complete knowledge of the various phenomena that are characteristic of an ice age. It is even possible that we are living today in an interglacial period, and that the earth will experience a new wave of glaciation within the next fifty thousand or hundred thousand years.

For a long time now, geologists, physicists, astronomers, and chemists have been seeking an answer to the question why ice ages develop. In principle, an ice age might be caused by a decrease of solar radiation of the earth's surface over hundreds of thousands or millions of years. Even a very small relative decrease of the influx of solar energy can result in glaciation, provided only that the decrease persists over a sufficiently long period. Imagine that the masses of snow which fall during winter do not melt completely during the next summer; the result will be a gradual growth of the polar caps, and of the glaciers and snow fields in the high mountain ranges.

At present, the opposite is happening. During the last fifty years, the polar caps have decreased to a noticeable degree, and the glaciers in the high mountain ranges have receded conspicuously. These changes have become apparent within the short period of half a century; it is altogether within the range of possibility that a new ice age may develop gradually over tens of thousands of years as the result of a diminishing energy influx from the sun.

The question is, what kind of natural processes could cause a variation of solar irradiation of the earth? There are several possibilities. First, there are processes going on within the sun. When observed over brief periods of time, the solar radiation is surprisingly constant. Even the eleven-year cycle of sunspots has but a negligible influence on the total energy emitted by the sun per unit of time. However, the sun has been observed closely for only about two hundred years, a period far too brief for accurate observation with regard to total variations in radiation that may have occurred during the past five hundred million years. The British astronomer Uno Öpik, who carried out theoretical studies on the physics of the sun, came to the conclusion that the sun undergoes periodic reorganizations in its interior which recur every two hundred and fifty million years, and which cause the total energy output of the sun to decrease in the same rhythm at brief intervals of a few million years. Such rhythmic variations would explain why the last three great periods of glaciation occurred at intervals of about two hundred and fifty million years. According to this hypothesis, we are presently in such a brief interim, a rare but most interesting phase in the life of the sun.

Another theory of ice ages is based on a variation of the carbon dioxide content of the earth's atmosphere. Carbon dioxide in the atmosphere acts like the window panes of a greenhouse, which allow the solar energy to enter in the form of visible light, without allowing the return of energy in the form of heat radiation in long wavelengths. For that reason, the interior of a greenhouse, or of a room with closed windows, is always much warmer than the outside. As soon as the carbon dioxide content of the atmosphere decreases, the mean temperature at the surface of the earth begins to decrease. A reduction of the average temperature at the surface of the earth by a few degrees would be sufficient to cause glaciation.

It is indeed possible that the ice ages are caused by a variation of the amount of carbon dioxide in the earth's atmosphere over geological times. Carbon dioxide is one of the main ingredients of the exhalations of volcanoes. A substantial cessation of volcanic activity in the course of the earth's history could therefore result in the development of an ice age.

Periodic perturbations of the earth's orbit around the sun could provide another possible explanation of the occurrence of ice ages. If the orbit of the earth gradually assumed a more and more elliptical shape, the earth would spend increasing intervals of time at larger distances from the sun. Even small changes in the shape of its orbit would result in considerable changes in solar radiation.

The Yugoslav climatologist Vilo Milancović reported interesting results of studies concerning perturbations of the earth's orbit as produced by the influence of the large planets during the past 600,000 years. The orbit changes caused by these perturbations coincide surprisingly well with the four subperiods of glaciation that occurred during this period. However, according to this theory, ice ages would have happened at about the same intervals during the entire history of the earth, a postulate which is not borne out by geological evidence. In all, the mechanism that caused the ice ages is still a puzzle to scientists.

All the present concepts concerning the development of the climate during the long history of our planet are still in the hypothetical state. It is certainly true that scientific research has provided brilliant results about many aspects of climatologic development; yet a satisfactory, comprehensive theory is still lacking. Geology and paleoclimatology will always suffer from the fact that intelligent beings were not at hand to witness and record the development of our planet over the billions of years of its existence.

Perhaps this is precisely why the study of the earth's history is so fascinating. In this study, imaginative speculations are not only permitted but most desirable. Present concepts are still based largely on speculation. However, a transition from speculation to confirmed theory has begun here and there, although none of the scientists presently engaged on this subject would be bold enough to say that he has grasped the sequence of events in the history of our planet. Nevertheless, there is the expectation that scientific research, in its continuous progress, will gradually approach a true understanding of the nature of the universe.

8 ✳ Future of the Earth

What does modern science have to say about the future development and fate of the earth? Any prediction, of course, can be no more than a speculation. However, speculations can nowadays be based on a large number of observations and exact measurements of the features of our planet, and may therefore be accepted as fairly reliable predictions.

Man has been present on earth for about one or two million years. This period of time is not much more than approximately one hundredth of one percent of the time the earth has been in existence. Thus he has witnessed only a minute interval of the earth's history. It is to be expected that the earth will outlast man by a similar factor. However, there are no scientific arguments against the assumption that the earth will continue to exist for several billion years to come.

Even if mankind does not commit worldwide genocide by means of atomic weapons, man's existence on earth cannot be expected to endure that length of time. A glance at the philogenetic cycles of animal and plant species during paleontological history reveals that most of these species enjoyed only a brief span of life. The saurians, for example, roamed the earth for less than a hundred million years, a very brief interlude in the history of the earth. Other species have survived longer—some of them are still extant, among which are not only primitive organisms such as protozoa and algae, but also higher forms such as cockroaches, scorpions, and turtles, which have been around for more than three hundred million years. There is no reason why these animal species should not persist another three hundred million years.

What brought about the extinction of certain species is difficult to determine. Several hypotheses have been offered to explain the relatively sudden disappearance of the saurians. Perhaps cataclysmic changes in climate occurred that had a detrimental effect on the environmental conditions of these animals; perhaps new species, such as the mammals, better equipped for survival, contributed to their extinction. The real reasons for their disappearance are unknown.

Because it is so difficult to know what caused the extinction of many species of plants and animals in the past, it should be no surprise that man's future cannot be told with any certainty. The uncertainty is further deepened by the fact that man, although a part of terrestrial life and subject to all the chief functions and laws of life, is distinctly different from all other living beings in several respects.

Man is endowed with reason. He has a perception of the essential processes in nature. It is not impossible that he may even succeed in recognizing and eliminating those forces that threaten his extinction. There is indeed a hope that the human race may enjoy a future that could endure millions of years. From today's perspective, however, man appears to be his own worst enemy. It is not unlikely that he may end in genocide long before he is overtaken by the natural forces that govern the evolution of the earth and of life.

Speculations of this kind are difficult to support with scientific argument. Other influences and forces are bound to contribute to the shaping of man's future. There is, for example, the relatively simple matter of how the development of the earth as a planet will continue in the near and more distant future. How much longer will the earth remain a habitable abode? How much longer will the conditions on its surface—climate, solar radiation, and all the other environmental factors—remain hospitable to life?

The earth, as was pointed out, is the only planet in the solar system that has oceans. The oceans represent probably the most precious asset with respect to the origin of life. They have been, and still are, of decisive significance for the favorable development of the climate and for the entire economy of life on earth. Without them, life as it now exists could not have developed. Whether—and how—life will continue to thrive, therefore, will depend primarily on whether the earth will be able to retain its oceans.

The oceans, being several billion years old, have thus far successfully survived all the changes and hazards during the evolution of the planet. On the other hand, a planet can lose its water content, as shown by Mars, where conditions are such that a large body of water cannot survive.

Even the earth loses a considerable amount of water every day. Once in the atmosphere, some water vapor may rise to high altitudes where it is exposed to intense solar ultraviolet radiation. Ultraviolet light quanta split water molecules into hydrogen and oxygen. Hydrogen escapes into space, whereas oxygen remains in the atmosphere. However, the earth has not really suffered from these continuous water losses in the course of its history. Actually, the amount of water within the oceans has been increasing during geological times. The atmosphere, too, will persist for a long time to come. Its most important constituent for animals and man is oxygen, but there need be no fear of losing the oxygen as long as plants exist.

A global war with atomic weapons could radioactively contaminate the surface of the earth to such an extent that not only animal but plant life would die out, in which case the oxygen in the atmosphere would also gradually disappear. After about three thousand years, it would have been consumed by the oxidation of rocks and minerals in the earth's crust, without any way to replenish it.

Again, as noted earlier, another important environmental factor on the earth is the radiation of the sun. Life on our planet depends absolutely upon the continuation of this uniform supply of energy. The fate of the earth, therefore, is closely tied to the fate of the sun.

Astrophysicists who have been studying the development and the life cycles of typical stars have gathered enough knowledge to make trustworthy predictions about the future of the sun. Although the sun may appear to hold an inexhaustible store of energy, its supply of fuel is by no means unlimited. The sun's fuel is hydrogen, which slowly is transformed into helium by nuclear fusion occurring in the extremely hot and dense core. Tremendous energy is thus released, emitted in all directions as solar radiation. A minute fraction of this radiative energy is intercepted by the earth, where it maintains life, weather, and other processes requiring a source of external energy.

At some time in the future, the hydrogen supply of the sun will begin to run low. It might be concluded that this will cause a slow but steady decrease of the intensity of solar radiation. However, this will not be the case. Paradoxically, a star increases its emission of light when its hydrogen supply runs low. The sun, in the far distant future, will not become hotter, but will substantially expand. In fact, the solar sphere may grow so much that it reaches the orbit of Mercury, causing this grossly inflated sun to hang in the earth's sky like a giant balloon, throwing off an amount of radiation that will strike on earth with a violence far above its present intensity.

The sun will reach this extreme state only in the far distant future—several billion years hence. It is not difficult to imagine what will happen when the increasing solar radiation gradually heats the earth. Although the present theories about future increases of solar radiation are not at all accurate, a prediction can be made based on a weighted average of the various theoretical estimates.

First, the oceans will evaporate; the water vapor, together with the atmosphere, will escape into space. Then, the temperature of the surface will rise until the surface materials melt and again assume the liquid state of eons past. Life on earth, of course, will have terminated by that time.

The next step in the development of the sun will occur suddenly, at a time when its interior has reached a critical state. Within a few days or weeks, the sun will change into an entirely different type of star, a white dwarf. It will shrink to the size of a small sphere no bigger than the earth, without losing appreciable mass. The sun will still be able to hold all its planets in their orbits. In fact, the planets, including the earth, will survive this cosmic catastrophe without any significant change in their orbital parameters.

As the sun is transformed from a red giant into a white dwarf, its total luminosity will again decrease substantially. It is quite possible that the sun will then continue to shine for some further billions of years, with a brightness much like the present.

These thoughts have stimulated a great deal of speculation. It is not unthinkable that at that time the curtain will rise again for a second act in the history of life on earth. When the solar radiation on earth has reached its former level, the earth's surface will cool off once more, its crust will resolidify, and volcanic activity will persist sufficiently to give our planet a new atmosphere—and even a new ocean. The new atmosphere and the new ocean may have a lifetime of several billion years, long enough to offer another chance for life to develop. Thus the fantastic story of life on earth might repeat itself.

As the white-dwarf phase of the sun draws to a close, the solar life cycle will approach its termination. The sun will gradually cool off, and in the very distant future the earth, orbiting around a dim sun, will be covered by a shield of ice and solid nitrogen. This is how scientists presently picture the final state of the earth.

All these events, if they occur at all, must be projected so far into the future that they are of academic interest only. They will not directly influence the fate of mankind. There is, however, the chance that, in the less distant future, other events and changes on earth will threaten mankind—for instance, natural catastrophes, such as floods or earthquakes. They occurred in the past, and they will surely occur in the future. However, such catastrophes are local in character and do not represent a real threat to mankind as a whole, and even less to all life on earth. Earthquakes and volcanic activity are closely related. Volcanic eruptions, as mentioned before, have been of great significance during the geological evolution of the earth. There would otherwise be no oceans. Earthquakes, therefore, should be considered as a necessary evil, as they are among the geophysical processes that the earth will never be without. Perhaps scientists will someday learn to predict them more accurately, and so help avoid the losses in human lives and property they often cause.

Earthquakes, hurricanes, and tidal waves are so fearsome because they strike almost without warning. Still other phenomena on the earth's surface cause catastrophes, though of less immediate danger because they arise from gradual processes. It is known, for example, that the level of the sea has been rising slowly ever since the end of the last ice age. The reason for this rise is that much of the ice that covered large areas on the earth a hundred thousand years ago has melted, and the melt water has caused the oceans to spill over. This will go on for the next fifty thousand years; as a result, the level of the oceans will rise by another hundred, two hundred, or even three hundred feet. Some of the largest centers of population, among them New York, Hong Kong, Hamburg, Los Angeles, and Rio de Janeiro, will eventually become submerged. However, it will be such a slow process that man can easily adapt himself to the changing environmental conditions, as in Holland, where the sea level has been rising above the sinking land during historic times. Even within the past few years the rise has been noticeable.

A new ice age would be another catastrophe of the slow kind. This could occur during the next fifty thousand to one hundred thousand years. It would prevent the further rise of the sea level, but it would cause a sharp worsening of the climate all over the earth, depriving mankind of large areas of arable land. In fact, an extensive glaciation of the planet would cause considerable lowering of the sea level, which in turn would uncover millions of square miles of new land in nonglaciated areas. These drastic changes would not have catastrophic consequences because mankind would have thousands of years in which to adapt to them.

One kind of natural danger, however, combines the two properties of a real catastrophe—sudden impact, within a few days or a few weeks, and global extent. Fortunately such events are extremely rare.

For many centuries, astronomers from time to time have observed the appearance of a new star, or nova, which could result in this kind of catastrophe. The designation is actually a misnomer, because the "new star" often proves to be an explosion of a star that had existed previously. Within a few hours or days, the luminosity of a nova increases by a factor of ten thousand to a hundred thousand, making it visible to the naked eye, while it may have been a dim star of low magnitude before the explosion. In a number of cases, by comparing new photographs with old photographic plates, astronomers were able to identify the new bright star as an inconspicuous little star that had previously existed in the same locality. A nova explosion is of relatively short duration. After a few years, its luminosity is reduced once more to its former value.

The future of our sun may hold a nova explosion—although the word "explosion" is not the most apt; actually, it would be a violent collapse, accompanied by the emission of enormous quantities of radiactive energy which would burn the earth to a crisp.

Besides normal nova events, there are the supernova explosions which are far less frequent, occurring once every three hundred years within our galaxy. A supernova increases its luminosity within a few hours by a factor as high as a hundred billion. For the duration of a few days, a supernova may be as bright as all the other stars of its galaxy added together.

During recorded human history, three supernovae have been observed in our own Milky Way: the supernova of 1054, which was described by Chinese observers; the "new star" discovered by the Danish astronomer Tycho Brahe (1546–1601) in 1572; and the super nova observed in 1604 by the German astronomer

Johannes Kepler (1571–1630). These new stars were so bright that they could be seen by day. However, they occurred in fairly distant regions of the Milky Way, and they appeared to the observer as bright but very remote stars.

What would happen if a neighboring star of our sun turned into a supernova? A star that was barely visible to the unaided eye would appear as bright as the sun to the observer on earth. A few weeks or months following such an explosion, the earth would be struck by a tremendous wave of cosmic radiation that would affect all living beings and their offspring in a decisive manner.

The dimensions of our Milky Way are so vast that on the average all supernova events occur too far from earth to present an imminent danger. Within a period of several hundred million years a supernova in the vicinity of the solar system could occur. It is not unlikely that a few such catastrophic supernova explosions in our vicinity may have taken place since the time when life began to develop on earth.

As early as 1945, I mentioned the possibility that supernovae near the sun could have been responsible for those puzzling discontinuities in the development of life that so far have defied explanation. Perhaps this is why many species of animals and plants vanished in the past, making room for new varieties. If the earth's surface were heated by a second sun appearing in the sky for a short time, storms of cataclysmic intensity would occur, with effects difficult to imagine. However despite enormous loss of people, animals, and vegetation, life would persist. Far more fearful would be the radiations.

A supernova explosion is not only a source of light and heat radiation; it also produces a gigantic wave of gamma rays, for example. It would be hard to tell whether most of the radiations would be absorbed in the atmosphere or whether they would penetrate it and endanger all organisms on land and sea. Such a shock would undoubtedly bring all kinds of biological mutations into existence.

Nature has given us a beautiful and fertile planet on which life was able to develop. A natural catastrophe capable of making this planet unfit for life—such as a star's exploding in the vicinity of the solar system—need not be expected even in the remote future because, as pointed out, the odds in an orderly universe are enormously against such an event. The only real and immediate danger to man lies in his newfound means of annihilating himself.

Index

About the Author

Heinz Haber was born in Mannheim, Germany, and received his Ph.D. in physics in 1939 and in astronomy in 1944 from the University of Berlin. He served as a research assistant at the Kaiser Wilhelm Institute of Physics from 1937 to 1939 and as chief of the Spectroscopy Department at the Kaiser Wilhelm Institute of Physical Chemistry from 1942 to 1945. Dr. Haber came to the United States in 1947 and was a research scientist with the U.S. Air Force School of Aviation Medicine, Randolph Field, Texas, from 1947 until 1952. In 1952 he joined the Institute of Transportation and Traffic Engineering of the University of California as associate physicist, and in 1956 became chief scientific consultant to Walt Disney Productions. At that time he was also made visiting professor of engineering at the University of California, Los Angeles. Dr. Haber is widely known here and abroad for his award-winning radio and television programs on science. In 1958–59 he conducted a weekly science program on CBS-KNXT-TV in Los Angeles and since that time has produced numerous science programs on radio and TV in Germany. Dr. Haber is the author of *Man in Space; Our Friend the Atom;* and *Stars, Men and Atoms,* which won the Thomas Alva Edison Award in 1962. He now makes his home in Austria.